CONNECTIONS
Toolkit for living in an energetic field of love

Copyright © Regina Orchard, 2016

All rights reserved. Without limiting the rights under copyright above, no part of this publication shall be reproduced, stored in or introduced into a retrieval system or transmitted in any form or by any means (electronic, mechanical, photocopying, recording or otherwise), without the prior permission of the copyright owner. In most cases permission with recognition will be freely granted.

Disclaimer

We care, but you're responsible.

This book is general in nature and not meant to replace any specific advice. Please be sure to take specialist advice before taking on any of the ideas.

Regina Orchard, her employees and contractors disclaim all and any liability to any persons whatsoever in respect of anything done by any person in reliance, whether in whole or in part, on this book.

First published in Melbourne, Australia by the author, 2016

ReginaOrchard.com

10 9 8 7 6 5 4 3 2

1st Edition, September, 2016

ISBN: 978-0-9954201-0-6

To my darlings
Merryn, Ethan, Adin,
Regina and Norm,
and all you are
connected to.

Contents

1 A field of love: what, why & how. 1
We all live in a field of love. 1
The challenge 2
You don't have to doubt your discoveries 4
Working with this book 4
Imagine IF 7
My story and your story... go for it! 7
Stand to access your intuition 11
EMERGENCY FIRST AID when it seems too hard 12
Homeplay 13

2 You've already proven it 14
People's emotions affect their own DNA..from afar. 17
Connection exercise 18
Your emotions affect my DNA, and vice versa 20
How you affect your own vibration and health 21
Energy and protection 26

3 Our connections create our world. 29
Astounding how a few people can affect many 29
From hopeless to full of hope 31
Offering a gift of harmony, not control 32
Cultural Creatives - You Are Not Alone 33
Creating and blocking connection 35
    Checklist for creating and blocking connection 35
Our beginnings 36

Dissolving obstacles to connection 37
Create your own map, choose your own ending 39

4 What you believe is true! 41

We create our lives 41
First focus on what you want in your life 42
We create our reality... the easiest ways 45
Brain watching 47
Brain washing 48
    Visualising 50
    Exercise: reflecting on my fabulous life 50
    Affirmations 52
What is most valuable for you to create? 54
Trying to believe you are the writer of your life? Needing more proof? 56
Happiness and Appreciation 58
As soon as you blame you give away your power .. 59
Can words deter you? 60
I have no power against people who are "doing bad" 61
What you believe is true so everybody is right! 63
These are true for you!! 64

5 Tune in and read the signs 65

Tuning In 65
Pain is a sign 66
Read more signs 68
Here is a message that most would miss 69

Things that happen to us!!!! 70

6 Caring for your physical self .. 73

Checklist: What does vibrant, abundant, good physical health look like? 73

To help, you need to know what went wrong 74

Grounding or Earthing 75

Each part of us affects every other part 76

Many of us do not even breathe well 77

I just woke up like this 78

When we carry more weight than we prefer 79

Hope for serious illnesses 82

    Much more than food 86

Action Checklist 1: Assist in balancing your health . 87

Action Checklist 2: Assist in balancing your health . 88

Dirty Dozen and Clean fifteen foods 89

    Here's the Dirty Dozen: 89

    Here's the Clean Fifteen 90

7 Caring for the rest of yourself 92

Your checklist for vibrant, abundant, good energy health 92

Energy 94

Our energy and emotions affect each other 98

10yo boy needing energy protection 100

Whose thoughts am I thinking? 101

Your checklist for vibrant, abundant, good mental health 104

Boundaries 105

Can I show Who I Really Am? 106

Benefits 106
Integrating my shadow 108
Feeling safe to be fully me 110
Competition... to be like someone else 114
Your checklist for vibrant, abundant, good consciousness health 117

8 When you block bad feelings ... 118

Your checklist for vibrant, abundant, good emotional health 118
Time bombs or allies? 119
Getting it right? 121
There are light and dark times for us all 121
Emotional Health 122
Whole Brain working together 123
"Being" in your heart or your head 125
Children's values 127
Your Inner Child 128
Ask a friend for what you need -for your inner child 131
Exercise - Your Young Adult Self 132
Write a letter to yourself from a different perspective 134

9 Feeling yucky is just a habit 135

Now for the good news 137
Appreciation or gratitude experiment 139
Checklist: from yucky to great 140

10 HELP, the world's unexpected. 141

Common questions from activists 141
More exciting experiments 142
The story so far: a summary 145

11 Hiding to thriving with others 149

Two People Loving 149
From I to We 150
When each of us show our hurt "inner child" or "pain body"
.... 152
Your Greatest Ally 154
I am the one 156
Other Allies 158
Nonviolent communication 160
Bargaining 164
From stress to creativity 165
Story sharing and tribe 166
Checklist: How are your relationships? 167

12 Playfulness in your life 168

Recipe for a fruitful life: add playfulness 169
Riches occur where play and intuition meet 171
You can change clouds 172
Are we playing with children? 173
Exploring playfulness 175

13 Creating our New Futures 181

Adding In and Fading Away 182
EXERCISE: Adding In and Fading Away. 182

A short exercise on NEWNESS 189
Visioning to make it so 190
Living in your desired life..................................... 191
Temporarily Lost on the Journey 191
When you're not the meditating kind 194
Small Steps ARE Powerful with Sue 195
Small steps ARE powerful Checklist 199
Living in your desired life..................................... 200

Bibliography 201
Acknowledgements 203
The Author 204
Where to from here? 205

1

A field of love: what, why and how

We all live in a field of love.

> *"We need to see ourselves as basic miracles"* - Virginia Satir

We all live in a field of love. Yes; you, me, everyone.

No, this is not romantic fiction. This is one of the big truths on which our lives are based. We are all literally connected, and that includes everyone and everything.

The hows and whys have been long taught and disputed. We have struggled to understand or believe. Have you ever felt conflicted when your experience disagrees with what you believe to be true?

To make this clearer, let me paint you a picture. Imagine a school full of cute little animated dolls. When they begin schooling they are whole, unique, and full of life. But by the end of their education each has changed, and the right side of their brain and body is missing.

It has happened gradually over the years, it has happened to everyone, so it's accepted as normal... Imagine seeing them all there. They have lost much of their full selves.

What if I now tell you that something like this has happened to most of us? Something as bizarre as this.

In real life our bodies and brains are still there of course, but in fact half our brains are neglected, and much or our wholeness is too.

At times in our lives there are strong prompts to alert us to this imbalance, and we will respond. However our education, what we have learned, can powerfully keep us from really exploring or claiming our wholeness. We deny we can be what now seems extraordinary or impossible to us.

What would it take to help you to see how extraordinary our world really is, and the potential that we each have?

Would it help if scientists from diverse fields had collaborated and, even better, had explained how we can regain our whole selves, and be more than most of us have ever imagined?

...Well you're in luck, because many scientists have, and I will be sharing simply what they've found and how you can use it.

The challenge

There are reasons for all the "good" things happening around us, and the "bad" things, and they are as real (in fact more real) than your own body.

The challenge for us is that we haven't known what to believe. We have been led by many opposing beliefs, dogmas and information.

I will touch on these briefly, and then begin with what might seem like magic. I will show you how to engage your virtual magic wand, and make new parts of your world appear.

Our own education system is very out-of-date about science, at least till the end of high school, and often beyond. So our education is based on old science. We favour specialisation, reductionism and separation in our areas of study. This has flowed over to our teaching focus.

These subjects tend to be dismissed as inconsequential:
- music
- art
- compassion
- emotion
- energy
- awareness
- use of our six senses

These subjects are the ones that activate and grow our right brains... our imagination, creativity and connection. By putting so little attention on these in schools, basically we are then invalidating whole brain connections, and blocking our bodies, hearts and minds from fully working together. We also invalidate the strength of our connections with other people and our natural world. If you're doubting this, please stay with me, there are deeper explanations, and new ideas for you throughout this book.

Just ask those who have dared to step out of "normality" and expand themselves to their true potential. Ask the scientists, who have dared to go beyond the limits. Ask the artists, healers, visionaries, and parents. There are some in all walks of life.

This book is a great place to start:
- If this is where you want to go
- If you'd love to realise that we now can do things that most of us could never dream before
- If you have an impulse to make a difference in the world, and shape it with more love
- If you want your uniqueness to be validated
- If you want stories of what is possible

(Of course there are many other places that will help you expand your life, and help your world, and I'll tell you about many of them.)

You don't have to doubt your discoveries

What I want to shout to you from the rooftops is YOU DON'T HAVE TO DOUBT YOUR DISCOVERIES. There are universal truths, now proven with scientific rigor, which back you up. (You'll find a couple of great ones in Chapter 2.)

Your weirdness, your woo-woo-ness, your insights, your knowing, your messages and unnerving coincidences, all these are real and valid. In fact there are enough discoveries to explore that would take you way beyond any strange possibilities that you can so far imagine.

Working with this book

This book explores how to simply apply some of the newest scientific discoveries into our lives.

I have chosen ones that can greatly enhance your life, and make your world a more loving and healthy place.

You will find proofs, stories, exercises and connectors to jump start some new options for you.

In the very best way, from working with this book, you will never be the same again, and our world will be better for it.

Here are some keys to help you begin using this book.

1. Start reading anywhere that appeals to you.
2. Start applying what you are drawn to... Every change you make will affect the rest of you.
3. No matter what you think of yourself: we are equals.
4. Kindness to yourself is always the key.
5. Start small, start slowly.

6. Begin by adding, not taking away.

7. It's never too late to move in a better direction.

8. Pain and discomfort of any kind are signs telling you to change.

9. Link with others for mutual support, ideas and inspiration.

10. Notice your intuition and let it guide you.

11. Begin to note the synchronicities (apparent coincidences) that come to you.

    I am certainly telling you that there are extraordinary possibilities available to all of us. But I'm not expecting that each of us who hears this will immediately become a master of this so that your life will transform instantly to a previously unrecognized state. Then again it IS possible! So if that's you, I will be so delighted to hear about your transformation. All of us are keen to know what others have achieved. We show each other what is possible, so we are opening doors of expansion for each other.

    For most of us this book is opening a new doorway. It is a basis to begin to expand our view and our life. This doorway may start as small as a crack in our current reality, and it can be some of the inspiration to move us forward and start to notice our intuition urging us to try new things.

    I am introducing many ideas here, and some of these may draw your attention. Perhaps it will be just one idea.

    And as you add this newness to your daily life, you begin moving in a new direction, and it makes the crack open wider. This step leads you to another addition, and the crack opens wider still. Eventually it could be as wide as half of your "house". And then you suddenly can see that you're actually inside a much larger mansion that you couldn't see before. This is your mansion, your domain of possibilities.

So start with what you are drawn to. Allow it in with as much ease as you can. Kindness to yourself is always the key.

Begin to note the synchronicities that come to you. Synchronicities are apparent coincidences that appear with ease, and at any time. You may feel like going somewhere or doing something, with no logical reason, and just the right person or thing is there at the right time. When you are in the web of life (and aren't blocking it) you are in "flow" and this is how it all plays out. (If you're not sure about this it will all be explained later.)

This means we don't have to try to plan all the details of our lives. Certain solutions occur without our conscious attention and in ways we haven't planned. And there is the pleasure of these surprises: a friend, a gift, a discovery, a large or small adventure.

Yes, I am saying that each of us has more potential and choices than we have so far imagined. And most likely it's FAR MORE potential.

"Potential for what?" you may be asking.

Tell me your dreams, your deepest urgings, and I'll tell you that they are all possible. In fact they are your soul speaking to you of what's possible.

What are the dreams of your soul?

Yes, you're right, one person can't make it all happen. But enough people can. And one person CAN make a difference. So you can go on your own adventure and...

Ok, I'm getting ahead of myself. This is only the intro, and there are a few chapters for you that will explain all this. Here are some of my dreams, and I believe this book helps show you a reality where they all are possible.

Imagine IF

- Each year your life gets better than the last one.
- You feel a loving connection to other people (even the crazy or mean ones).
- You can see how your intentions work for you and for others.. BONUS!!
- There is more world peace.
- There is more generosity.
- We have a sustainable and healthy world.
- There is a real future for your children and their children.
- This is a good world for animals and plants too.

My story and your story... go for it!

Let me tell you more about this, about making it much easier to live the life you desire.

It's not something I have lectured on back over the last big chunk of my life, because I didn't know all this. Or to be more accurate I did know a large part of it - intellectually - and THAT was part of the reason I haven't truly believed it.

And then some unexpected things happened to me over the last few years, and it has opened my eyes to somethings that never occurred to me before.

Let me start by telling you one of the stories that began all this, a long time ago in my life. In the time before I was a mum, I was in a sales and management program. My whole income was commission, not retainer, and I'd been working at this job for a year or so. Now I'd never had a powerful sales persona. I'd left school so shy that whenever I walked into the large cafeteria at my first job, I felt the imagined pressure of every eye scrutinising me.

So ten years or more later I was still fairly shy. I knew my sales script inside out, and deftly delivered it, tailored to each of the hundreds of people that I "sold". But every time I reached the simple "closing questions" at the end - the equivalent of "Would you like that in blue, red or black?" Or "How would you like to pay for that... cash or credit card?" Each time I was still shaking inside... nervous rather than comfortable.

Despite this stress I was earning a living with sales, and I was enjoying having a new man in my life, a fellow salesperson.

One morning at work, wearing my cork platform sandals about 7cm (3") high, I hugged that man, happily excited, and said these words, several times: "This is so good I CAN'T STAND IT".

Almost immediately, still hugging, I fell off one of my shoes and twisted my ankle. I did so much damage that I could do very little work that week. I couldn't knock on doors, and spent much of the days at home with my foot up.

There is much power in our emotion and the intention we express. A high emotion, positive though that is, is a great force for creating in our world, and I really did keep saying that "I can't stand it!".

You may be surprised to hear that I still managed to be salesperson of the week. Here's how it happened.

One of the things that I DID do that week was read a small book of affirmations: "Your Word is Your Wand" by Florence Scovel Shinn. I read them, I affirmed them. You could say that I spent quite a lot of the week mindfully expecting good and abundant results in my life (see Brain Washing).

At the time I definitely attributed the powerful sales results - my best results to date - to that affirmative and contemplative time.

Many times since, I have felt my behaviours created results: sometimes what I desired, and sometimes what I was trying to avoid. To learn more about this I've intentionally explored various systems, processes and beliefs.

But science has always been my friend. My first two jobs were data management and computer programming. Both were predicated on the importance and reliability of science and language, comprising sequential, logical thinking. In short, my education had led me to use the logical part of my brain that emphasises and values itself (and ignore or be overwhelmed by the rest).

I think I was a typical and even so-called "successful" product of education - though I haven't felt particularly "successful".

I was not well versed in how to use my sometimes powerful emotions. I was certainly not educated to value or use my intuition. I also was discouraged from thinking that I could make a living from any creative endeavour. Then there was dancing, movement, art or music, love of good design or brainstorming, all of which I loved with much joy, but for which I didn't show any professional promise. Being capable with maths, science and English, I wasn't called "creative". So it never occurred to me that such options could be open to me, even though my interests were varied, including dance, tai chi, yoga, good design systems, sustainability, bicycling, community, clowning and communication.

Much later I realised that all of my non-career interests are attributed to either right brain or whole brain creativity. After integrating these, and intention and intuition, also known as gut instinct, into my life, I have had these changes:

1. I find life easier.

2. I feel more whole.

3. I feel more joy and satisfaction.

4. I feel more connected than I ever have before.

5. I often feel that I am full of love, and connected to our planet's nature, and to other people.

6. Things that I wish to have or do or be often happen. Often quickly.

This was a break-through for me. Many people have found these "secrets" and applied them, but many others haven't. I am excited, so I wish to share some of these universal rules with you, and point you to everyday tools.

Let me explain here that when I say, for example, "I feel full of love", that is something measurable[1] in my body. There is equipment that is used to monitor these changes in detail. My heart energy, or heart chakra, is stronger, and I can feel pleasurable sensations coming from that area and spreading into my body, as well as in my consciousness.

Having learned from many people, both personally and virtually, I've applied these concepts to see if they work for me.

Did you grow up like me, and find you had abilities or understandings which you doubted because outdated science invalidated them?

I doubted my own discoveries for many years. So I did a step forward, and sometimes many backwards or sideways, before I'd take another forward step.

---

[1] Measurable by Biofeedback and Kirlian photography. See Institute of HeartMath.

Stand to access your intuition

Here's a tool to learn and start using as you read on. To bypass the confusion of your mind, whenever possible use a method that accesses your body knowledge.

It's a simple exercise, please give it a go now.

1. Stand up, balance your weight evenly on both feet.
2. Relax your ankles and knees.
3. Close your eyes if you'd like.
4. Become very aware, or "tune in" to your body.
5. Ask a question. Usually a yes response is rocking forward, and a no response is a rocking back, but test it.

Note: The wording of your questions is important. Some can mislead you.

- Avoid "should" and "must".
- Be specific.
- Ask a simple question, not a complex one. Sometimes a question carries other assumptions, so it's not possible to clearly separate the parts and your answer can mislead you.
- Ask "Is this be the highest good for all concerned?" or similar.

Have you tried it? Did it work for you? Please try to use it daily for a few days.

Thank you to Karen from Sacred Soul Healing for this exercise, and for much of my understanding of working with energy chakras and applying the understandings in life.

Other options: Kinesiology and Expressive Arts processes can enable detailed conversations with your body - with your embodied wisdom.

EMERGENCY FIRST AID
when it all seems too hard

　　　Some of my shifts have been battles, and like everyone else I have tears and frustrations along the way, as well as the joys. At times when it all seems too hard, (and trust me, you will have some of those moments or days) please do one of these:

1. Flock with others... for support and ideas... and don't be shy about it, they all have tough times too, they'll understand.

2. Apply one thing[2] to help right now... you'll feel better when you are being kind and supportive to yourself, or are moving in your preferred direction.

3. Avoid general bad news media... No, it's not representing reality, it is focused on the worst in human life.

4. Move your body... shake, dance, walk.

5. Be outdoors / in nature / barefoot when possible.

6. Use one of your strengths in a way to help one of your weaknesses.[3]

　　　As a great ending to this introduction, it's good to make this a playbook (I'm using "play" instead of "work", to remind you to stay light hearted.)

---

[2] See Checklists and Exercises and Creating our New Futures

[3] The Wellness Workbook, Dr John Travis and Regina Ryan, and their Wellness Checklist

Homeplay

So how about taking some copies of this emergency list? Phone cameras are a help, or if you're reading this as an e-book take a screen shot.

Now where to put these copies? Where might you be next time it all seems too hard? Where will you need a reminder?

And as a preventive, will you schedule one of those 6 items in your calendar this week? Which number will it be?

Is it feeling good to take action and plan?

All the best, see you soon, in another part of the book.

2

You've already proven it to yourself, but you didn't realise.

This is almost the longest chapter in the book, covering a lot of topics, to give you an overview.

I'm asking you to find a way to stay with me by doing one of two things: read the whole thing; or if it seems too long, go to one of the other more specific chapters or sections that catches your eye.

This chapter has stories, experiences and some extraordinary discoveries about your world. But mostly it's about the impact that it's had on improving people's lives. So read on and see if you can identify yourself here.

[Throughout the book, to assist with conversations, my speech will be in plain text and others will be in italics.]

This chapter topic has arisen as significant, as so many people doubt the realities that occur to them. They think that faith is required to believe what even science now has proven. And that's understandable because our culture is running on guidelines from a past age. However each of us have had experiences which do verify these newer understandings. There are many things we have already shown to be true our lives. Let's see some of these for Isabelle and Zena.

My opening words were: "This topic is *You've already proven it to yourself but you didn't realise.* I'm curious to ask, does that have any particular meaning for you?"

I was running a session with Isabelle and Zena, two capable professional women, and Zena's response was not what you may expect.

"I haven't thought a lot about it, but it probably does. If I stop and think about it now, I tend to downplay what I achieve. I do counselling and I find that with my clients, the ones who I recognise as having similarities to me, they tend to do the same thing."

"So you downplay what you achieve?" I said, "Could you give me one example of that? Is there one that springs to mind?"

Zena: "I think it's more this tendency: although I would achieve something, I would think it wasn't good enough. I wouldn't stop and recognise or accept that I had done something!! "

"A very long time ago I joined the JCs, the Junior Chamber of Commerce (because my boyfriend was in it at that time). Very quickly I was recommended for the new JC of the year. Looking back, that was pretty amazing. But at the time, in my mind it was nothing. And really I should have thought "this is great, people are nominating me," rather than "... they just need someone to nominate"."

"I have found that's a common way that people respond Zena. Although you have proven your abilities to yourself, you still didn't realise; you didn't give yourself credit. That's what you're saying, isn't it?"

"Yes, not acknowledging my abilities! Not stopping and thinking that it was a magnificent thing." She paused again to consider, looking puzzled, and began to speak as if talking about someone other than herself. "Yes it's downplaying your abilities or not thinking it's good enough even if it really is. And if it was somebody else you'd be telling them it was wonderful. I guess that's the thing, if I think of it as someone else then I change how I feel about it."

Isabelle joined in, "I sure like the topic, it's very inspirational. Something draws me to the title." She was still reaching for the words that captured her feeling. "It's awesome."

R: "That's great. Is it more a general feeling, or is there anything specific in that?"

Isabelle: *"Let me think a minute. I have a service dog, and every day I think "I'm not qualified to keep on training this dog! I need the help of a professional." But I've had all the training, and I AM doing it! So I have been proving it to myself every day. Is that what you mean?"*

R: "Yes, that's it. Zena had a similar kind of meaning to it. My intention was slightly different, but it's interesting how it all fits together."

My proposition is basically that there are universal laws that science is now catching up with. It is now possible to prove some of these laws through Scientific Method. Basically many of us have some inner doubt, such as, "Well I know the universe works like this, but it doesn't work for me. Other people can create their life the way they want to, but I have problems with this and that, and it won't work." We believe that we have proven that things can't work.

But my contention is that there are many times in each of our lives that we've proven that we can be what you might call "being in the flow of life"... and things occur the way we want them to.

And on the other hand there's been many times when we have consciously proven how we can BLOCK things by the ways we think, feel or believe. Or by the way we don't see things. And so on. So my contention is that we have proven a lot of these things as well.

In all ways we have proven that we have control of our lives, except that it's usually not conscious, and may not be the result we intended or hoped for.

Later in the book we will be looking at another universal truth that this represents, "what you believe is true". But now I'm going to explain two scientific experiments which relate to this area, and do a couple of exercises with you.

People's emotions affect their own DNA... at a distance

The first experiment is about people's emotions affecting their own DNA. This experiment was done back in 1993 and was led by Dr Cleve Backster for the U.S. military. The first set of experiments was to separate people (yards) away. They precisely monitored the DNA and the volunteers, while they were showing them a series of very emotional video images. When all the data was matched, the results were astounding. Each time there were emotional peaks and troughs in a person, their DNA responded. As if this wasn't extraordinary enough, it occurred instantly, so with no time lag.

Dr Backster then took it further, repeating this same experiment over hundreds of kilometres away and the same instant response occurred. They acted as though the DNA was still connected to its owner. Even when the cells were hundreds of miles away from the person: the cells still reacted, at the same instant, just as they would if they were still in the body.

Zena and Isabelle were almost speechless as they tried to make sense of how the experiment's strange results could be possible. *"That's incredible." "So amazing!"* They sat listening again, waiting for some explanation.

Perhaps they wondered why they could be educated adults who had never had access to this life-changing information before. Yes, I had long wondered that so few of us know this... and that's why this book has urged me to write it. But back to the meanings for us from this experiment... Yes it's amazing. Yes and that then informs so many scientific areas.

Many scientists have experimented, postulating that the person and their DNA are linked by what is like a web that connects everything. And we are all connected into this web at all times. It's the only way many scientific observations and discoveries can be explained. It's their best explanation so far. And this is also the explanation of many ancient civilisations, such as the Native

Americans, and the Australian Aboriginals, and similar to many religions.

Here is the second major impact of this: Human emotions have a direct influence on DNA, on cells, on life. And distance appears to be of no consequence to the effect.

Connection exercise

Our first exercise is about connection. Please make sure you have something to write with, you might want to jot down a few words as things come to you while you read on.

Please first relax and take a few deep, slow breaths...

Now connection... Really think about or allow times to pop up from your memory.

Was there a time when you really knew who would be on the other end of the phone, when the phone rang? Or perhaps you picked up the phone and tried to ring somebody and they were actually in the process of ringing you?

Or maybe when you felt the energy in the room, like the saying "there was so much anger in the room I felt I could cut it with a knife". Or you walk into a room and it just feels wonderful there, or some other feelings or sensations.

Or a time when you had a kind of urgent warning. Or you had some kind of insight about a person, or a situation.

All of these things are quite different from our usual, logical view of the world.

As the memories pop up please make a brief record of them, and then relax and see if any more arrive.

.. Keep breathing as you allow enough memories to come to you.

So who has something to share please?

Zena: *"I can think of a few. I had a friend years ago I worked with at Uni, and it would happen constantly that I would think of her and she would ring me or I would know she was about to call. One time I remember having a daydream about her. I had to pick her up, her car had broken down. And I thought "that's ridiculous" because we lived in opposite directions, and there's no way that I would be taking her to come to my place. My day dream didn't make sense. But when I rang her it turned out that her car had broken down. So I made it different in my daydream, but the main thing was that her car **had** broken down."*

R: *"Yes they are instances of something other than our usual beliefs of how the world works."*

Zena: *"And another thing is with my children, I can tune in and see how they are, and that works I know. One time when my daughter was in physical danger, she was being attacked, I knew. She was meant to be somewhere safe, with another family. But I knew something was wrong. I just had this dreadful feeling the whole time. I actually wanted to get in a car and drive there, but I didn't know where they were camping."*

R: *"You've shared some very clear and powerful examples Zena thank you. How about you Isabelle, have you had experiences that you'd like to share?"*

Isabelle: *"I've definitely had experiences similar to that, as a daughter. Having traumatic experiences happen to me and having my mum know and call."*

"And with my friends, we have those shared connections. It happens to me all the time. I kinda know when people are thinking about me. I definitely notice that thoughts have some sort of real energy."

R: *"Yes they do. That's a great lead in to the second experiment."*

Toolkit for living in an *energetic field* of love

Your emotions affect my DNA, and vice versa

This group of experiments, all along the same line, was run between 1992 and 1995 by the HeartMath Institute.

They were about a person's emotions affecting another person's DNA. The volunteers were trained in what they called emotional coherence: to calm their minds and to focus on loving compassion coming from their heart while relaxing and rhythmically breathing.

They focused on human DNA samples in beakers, and two different intentions each caused a different effect[4]. To briefly explain: if you stretched out one DNA strand, in each of your trillion cells, it would be 3 feet long (almost 1 metre), so usually they are very tightly packed in the cell. They have to be unwound enough for the genetic directions in them to be read by various chemicals, etc, and then wound up back into place. This happens all the time.

In the experiment the DNA were monitored both visually and chemically for differences. It was found that both winding and unwinding of the DNA were being intentionally caused by this emotional/vibrational focus on the DNA.

Previously scientists thought that the only things that could wind and unwind our DNA were drugs, chemicals, and Electrical Fields. So it was quite startling to see how much these people could affect DNA, which is what life is made from.

These and other experiments all confirm the concept that we are in a participatory universe. So not only are we all connected, but all of us are creators of our universe.

We affect other people and life forms.

---

[4] For further research notes on this topic, including the mentioned research:
http://www.item-bioenergy.com/infocenter/ConsciousIntentiononDNA.pdf

Of course when you consider that everything in our world, including inanimate objects, are all made from similar particles, isn't it possible that we are connected to our homes, cars, possessions and rocks in our earth too?

There is a web that underpins everything.

How you affect your own vibration and health

So my question about this is not how you have affected others, because I'm sure you would have endless stories. But how are you aware of how you can affect and change your own vibration? ...That's your own state of mind or being... If you think of anything to jot down while you're reading, please do.

It will be helpful to stop and relax... focus on your body as you take 3 slow, deep breaths...

Remember a time that through your intentions, your actions, or your focus, you have changed the energy in your body. Did you lighten your mood? Did you feel more energized?

Have you intentionally used various tools to achieve various changes in yourself? Perhaps movement? Music? Things you read? Things you do?

Take some time to breathe, relax, and make notes...

And when you're ready, if one if you would like to share.

Isabelle: *"I have been aware of HeartMath. There's one of their books that talks about where you focus energy on your heart, and think positive thoughts towards and about things. It affects not only yourself, but also others. So I try and do that with meditation, and thinking about words that positively affect things. I would normally not talk about it. But I focus on words such as love, appreciation, inspiration and calm. Words are so powerful, and if I combine that with focusing on my heart it helps me, but I also believe it helps things around me. Sort of a blessing... I don't know."*

R: "Yes that's what science very clearly is telling us. Traditionally people have been living their lives based on these truths. Well science now has the finesse to prove that to people like me, who until now have believed in the limited old science that we've been taught."

Zena: *"I have a few different examples coming to my mind. Well I have done a few things that might be called "way out". I've actually trained in energy healing at a time when I wasn't particularly well. Mainly because my mother, of all people, recommended this energy healer as someone whom she felt would be helpful. Purely for financial reasons, after seeing her a few times I thought, 'This is so expensive I'm going to go and do a weekend workshop, because I'll get a whole lot of this at one time'."*

"The first workshop was after an operation where I was left with tingling and numbness down my left side. It was tingling constantly, really interrupting me in a lot of things. In this workshop was a technique which was meant to change your DNA by doing a particular type of meditation that an American meditator had come up with. And I just had the thought "this will fix it". I did it with my partner, who was focussing on me... and it's funny I haven't ever told anybody else this yet... I just felt this amazing sensation... I felt almost like I could feel a light, which sounds ridiculous, going down that part of my body. I had this image of a really white light going down that part of my body... Afterwards that part of my body felt different and I didn't get that symptom again. It just went away. That to me was something I wouldn't tell a lot of people I know, because they would just think I was wacky.

I have met a lot of people in the different workshops where I have trained, probably more than a dozen who have said to me "I have this illness, my doctor tells me I'm the only person in the world with this condition who has gone into remission. I am just a fluke." But there are so many of them that I have met in this situation! So I said to myself, 'this works!!' So that's had a very powerful effect on me."

R: "I am so appreciative that you both have things that you don't tell people but you feel comfortable to share here. And it's fascinating to me how many people are exploring this part of life that many people don't recognise yet. They are tapping into this side of living, which to me is really the main part of living. So really I guess it's not too surprising at all... But it's so powerful."

Isabelle: *"Have both of you heard of Dr Emoto's water experiment?[5] That was truly wonderful. (...Yes we have.) I'm quite blown away by the discussion. I really appreciate all that."*

Zena: *"Isabelle, I also have the Institute of HeartMath's book. The meditation, breathing into your heart, I find is very powerful.*

I know we're supposed to talk about ourselves here, but I actually teach that to my clients. I teach them how to breathe into the heart. And what I say when I teach them is "close your eyes" and I teach them the method, and I tell them 'just try it for 10 seconds and see how it works for you... 10 seconds is nothing', And invariably people start to feel very relaxed. You can feel their energy change when they do it. I think it's a very powerful technique."

R: "Yes, I have done it too, and just doing it again now as you talk I can feel how transformative it is... But it still seems extraordinary to have these responses in only 10 seconds."

Zena: *"10 seconds is really not very long to try is it? And I've only had one person who didn't notice anything in 10 seconds, and I've taught it to a lot of people."*

Isabelle: *"Similar to Zena I teach yoga once a week at my workplace in a pretty big building downtown. A lot of people know each other and me. So at the end I have people look into their heart. You might suggest that people incorporate this more into their daily activities. Because I find when I bring it into my day somewhere it helps in other areas. When I'm going into a meeting at work and I focus on my heart, or do some sort of thing that brings that energy, the meeting feels better."*

---

[5] Dr Masaru Emoto has numerous books on his discoveries, including The Hidden Messages in Water and Messages from Water and the Universe. He showed that thoughts and intentions changed the shape of water crystals as well as the quality of water. And remember how much of humans are water.

R: "I appreciate all the feedback. You and Zena have both had the opportunities to share the heart breathing in your work. Usually though, as you say, most of us don't talk about it. We just do it. We do all these things ourselves, but we rarely talk about it with others. So then we often feel we're unusual, and really there are many others like us."

Isabelle: *"I'm a business analyst. We have the workout place in our building, and I am able to say it out loud there when everyone's just done yoga....but I'm never able to say it in a meeting. So it's fun to have found a way to bring it to my workplace."*

R: "Yes it would be interesting for you if one day the right person comes to class who you work with, so that things change in the workplace, a bit more consciously."

Isabelle: *"Yes, that's been needed for a long time. I've been trying to figure out a way to heal my workplace for some people. I've done a lot of things to bring a bit of heart, such as bringing my service dog to visit."*

R: "As you've been telling me about yourselves, I should tell you there's been times in my life when I've used all sorts of aware and intuitive techniques when massaging and healing. And then at times I'd go through a kind of period where I'd basically deny it, as though it couldn't really have happened, I must have been tricking myself. And then one day I'd return to it again like a spiral, and I'd know it's all real. I knew it was all true, and then I used it some more. But it's been interesting how I have blocked myself at times."

Zena: *"I can relate to what you're saying Regina. I've had friends where medical science couldn't do anything for them, but they are drawn to something.*

For instance, Chinese medicine traditionally heals fluid around your heart. I know two people who've been successfully treated. It's a standard treatment - they can do it in 3 months. But it's not an option you will hear from a Western doctor. There are so many options around, it's just finding the right one for ourselves at the time. That's the hard part."

R: "Yes, their intuition led to solutions. I also have seen this happen for many people."

Zena: *"I think the other part of that was not just intuition, but somehow we set an intention and it comes to us. As I think back, 3 different sources told me about something and I thought "I need to pay attention to this". Because out of the blue someone would tell me about it, not necessarily trying to get me to use it, but they would just tell me about it. So things come to us if we are open to listen to it. And I'm not always the best at that."* (Zena giggles warmly)

R: "What you were saying about Chinese medicine, there is a video on YouTube, Gregg Braden has showed it many times. In a Chinese hospital three practitioners were basically meditating or praying - whatever you want to call it - and they brought a woman back to health. Her tumour, shown on the sonogram, disappears within 3 minutes while they're doing this. The Chinese practitioners in their hospitals have all sorts of knowledge that we know nothing about."

Zena: *"I would love to see that video if you have a link to it at any time."*

R: "I will chase it up and put that on my website: www.ReginaOrchard.com."

Zena: *"When I was younger I had a supposedly incurable disease[6] but my mother didn't tell me that it was incurable, and then I didn't have it when I got older. I believe our thinking makes a big difference sometimes too."*

R: "Yes, when we are told something about our health by a trusted authority, this can have a big impact on us. As your mother didn't tell you the diagnosis, you didn't have to fight against your doctor's opinion."

---

[6] The specific type of disease has been removed for the sake of the person's anonymity.

Zena: "I've also seen that remarkably with a woman I've counselled, with a terminal illness. When I saw her she was told that she had 24 months to live. She came in for anxiety because she was having panic attacks as she was certain she was going to die. When I stopped seeing her about 6 months later, she was off the palliative care list."

"There were a couple of really interesting things I noticed with her. First: that her new doctor had told her 'you've got 6 months and nothing can be done'. And as she had beaten this issue before, I encouraged her to see her previous doctor. And finally she and her husband travelled to see this doctor and his view was a total opposite to their new doctor. He said, 'Oh you'll still be around in 10 years. Don't worry about it too much, you've beaten it before, so no problem'."

Energy and protection

"And the other thing also is that I taught her a different meditation which is about putting protection around you. There's a whole lot of very similar methods where you have energy or love around you and you breathe it in, you have a force field around you and you breathe out all the bad things that goes through it. The bad things vary depending who you're talking to - whether it's pain or anger or worries. She took that on board and she said "I have a shield around me and I take that everywhere with me. And once a day I sit down and put my shield around me and make sure that it's there." And she and her husband both said that it's made such a difference to how she is now."

R: "Yes that's wonderful. I've done something like that shield for a long time too. More recently I've discovered that I need to remember more often to do it because I've been taking on people's energy all my life. It can be really exhausting and draining. It's so wonderful that she learned just to have her own energy and her own strength."

Zena: "Well I learned in an energy workshop to put protection around yourself every day, and you know I forget to do it. I've taught my kids that if they're ever upset to put the energy around them, and sometimes I help them put it around them, even over the phone, and they'll feel better. I really have to do more of that myself, I do start doing that, and I get too busy and forget."

R: "Yes I forget too. I've been known to get out my water colours and paint messages on my bathroom mirror to remind me... I forget all this wonderful wisdom if I don't have a system or something to help me do it. I love doing things such as exercising or energy work with friends, to make sure they happen, and I use my phone planner or checklists too. I currently have a daily reminder on my phone schedule to "Bubble Up & Shine" - that's my energy protection bubble and how I feel when I do it."

Zena: "I'm using other people's rooms for my clients but interestingly I found one that has really nice energy. Even a client came in and said "this is a much nicer feeling place than the last one you were in." It's interesting that she noticed it. She's not into energy stuff at all. She's very conservative but she noticed it enough to comment."

At this point Zena divulges a serious health condition she has.

Often it is easier to see how another person can be helped than ourselves. So my question to Zena is "If you had no restrictions on what you could say to your client, and your client presented with this health condition, what would you want to say to her?"

Zena: "I guess go back into really doing more things that help my energy. Doing protection. Doing healing. I know how to clear energy in places, so clearing energy in my home. I've been told there is a real weakness in my body, and I have this feeling that I need to find a way of meditating or focusing on how to heal that. There is a sense I have that there is a way of healing it, you know as long as I don't speak to anybody who's going to put me right down, because then I'll go into questioning my certainty. So I'm careful who I talk to. But the thought that I have is "I really need to look at myself first, look at healing myself, and not accept what I am told"."

Zena, Isabelle and I have recalled times here where in diverse situations we, and people we know, have all "proven it to ourselves". In many facets of our life we use these metaphysical (beyond physical) abilities and awareness. We are aware of consciousness, connections, energy and palpable experiences beyond our own bodies.

If you've recalled your own similar experiences, then this is already real for you, and you're ready to explore along with us in this book.

If it's not yet familiar to you, but you feel curious or as though something is awakening for you, please come along for the read, and see where this will take you.

Remember, life is a journey for us all, and sometimes all we need is to be willing to take a step in a new direction, and a whole new vista opens up, and with it the gifts of new experiences and understandings.

3
Our connections create our world

Astounding how a few people can affect many

> *"Meditation does not give you as much as you lose...you gain peace and focus, and you lose stress, fear, depression, anger, worry, resentment and panic."*

This is about the small number of people who can affect whole populations for the better.

Much testing has been done on the powerful effects of meditation. Princeton and Harvard Universities, with meditators from Maharishi University, did many experiments in the 1970s and 1980s. Certain cities were monitored while select meditators were using the equivalent of the HeartMath emotional coherence approach... stilling the mind and sending unconditional love, or compassion, from the heart.

There were observed[7] reductions in statistics of traffic accidents, fires, hospital cases and even crimes in these cities while the meditation was occurring. As well as this, during the Lebanon war, in 1982:

- war-related fatalities decreased by 71%
- war-related injuries fell by 68%
- the level of conflict dropped by 48%
- cooperation among antagonists increased by 66%

---

[7] - http://www.worldpeacesolutions.net/2015/12/05/a-scientific-solution-to-terrorism-and-conflict/

- International Peace Project in the Middle East, The Effects of the Maharishi Technology of the Unified Field: http://jcr.sagepub.com/content/32/4/776.abstract

- https://issuu.com/maharishiuniversity/docs/yogicflying1

Many times these experiments were repeated, to check that other factors weren't the cause of these clear trends. The footnoted New York Times article states, "These findings have been replicated in more than 50 studies, published in leading peer-reviewed scientific journals, and endorsed by hundreds of leading scientists and scholars. The efficacy of this approach is scientifically beyond question."

The mathematics was also done, and the "tipping point" where enough meditators made these changes, was found to be the square root of 1% of the population - $\sqrt{x/100}$, where x is the population.

CAPITAL CITY	POPULATION	$\sqrt{x/100}$	COUNTRY	POPULATION	$\sqrt{x/100}$
Beijing	20,180,000	449	China	1,339,190,000	3,659
Washington, D.C.	601,723	78	USA	309,975,000	1,761
Ottawa	898,150	95	Canada	34,207,000	585
Canberra	354,644	60	Australia	23,490,000	485
Santiago	5,084,038	225	Chile	17,114,000	414
Phnom Penh	2,011,725	142	Cambodia	13,396,000	366

The most profound thing in the table above: you can see how few people are needed to affect the consciousness of these cities and countries.

From hopeless to full of hope

This is the most exciting and heart-warming news I have heard in a long time. It shows that a small number of people co-operating for good outcomes can make huge changes in our world. It goes beyond any arbitrary borders of country, language, religion, income or education.

I wanted first to emphasise the importance of this, and now I will explain more about it.

As you can see from the numbers in the tables, it takes a surprisingly small number of people to make these changes. China's population of 1.3 billion people in theory can be vibrationally affected by less than 4,000 people. This is a tiny number of people to make a change.

In actuality it may need to be done on a more local level, but looking at the example figures in the table above, even for the largest cities it's less than 500 people needed, and under 100 for three of the cities. A few dedicated groups working together can potentially make significant effects in their town.

Notice also that China's population is 57 times larger than Australia's, while the tipping point is only 7.5 times larger. Similarly comparing USA and Cambodia, USA's population is 23 times larger, and the tipping point is only 5 times larger.

It is possible because we are all connected. Literally.

Offering a gift of harmony, not control

It's important to know that the meditators did not MAKE people behave differently. There wasn't one group of people forcing others. They were creating an effect that others enjoyed. They were holding out an offer of better vibrations, or a higher consciousness. It's much like you walk into a room and it may feel great, comfortable, welcoming, or perhaps quite the opposite with much tension and discomfort. Some people will be attracted by one and some by the other. In this experiment, obviously many people responded to this better offer. It brought out a better side of them.

So our emotions and awareness particularly affect our own DNA and that of others. But don't worry about people using mind control on others. This reminds me of a story of 2 thieves who hear that yogis can walk through walls. So their plan is to join the yogis to learn to do this, hiding their intent to use it for thieving. The yogi master is sensitive to their plan, but welcomes them. The studies take years, and of course by the time that they have the ability to walk through walls, they have gained so much in themselves and the joys of living, that stealing from others no longer appeals to them.

So by the time a person has developed their power, they are also connected... Truly empowered. A person in this state is not drawn to use power over others. This person helps as one of the people to create a better world (in the group of the square root of 1%).

The most powerful force is unconditional love or joy. We are told this by scientists, by natural healers and ancient cultures. Talk about intelligent design!! This is totally survival oriented.

Fortunately the most powerful force is NOT hate. Nor is it fear, which is the underlying cause of hate.

Also it's not violence, greed or coercion. It's love and compassion. Pure love which warms the heart and makes you smile for no particular reason. It drives you to want to share the good feelings. And they do rub off.

I realise this may sound very naive, but I will be explaining further how this is so.

Cultural Creatives - You Are Not Alone

About 10 years ago I made a great discovery when I found this book: The Cultural Creatives: How 50 Million People Are Changing the World.[8]

Everything I saw, from the dust jacket on, excited me. I wanted to know more... Who are these people Changing My World...? This could be great... I wonder what they're doing... That is a big group, am I one of them?

Yes I am one of them, and as you're reading this it's likely you are too.

One of the most astounding features of Cultural Creatives is they tend to all think **that they are the only ones like this, except for a few friends.**

They think they're alone in their world view. They aren't aware that approximately 50 million adult Americans (slightly over one quarter of the adult population), and at least 200 million worldwide, can now be identified as belonging to this group.

It is a huge change of direction in our world, and these people are creating ways to live which herald in a new era.

Do not feel you are alone. The world is full of creative and compassionate people putting their attention on building the world of their dreams. And yes, there are enough people to affect our world in wonderful ways.

---

[8] The Cultural Creatives: How 50 Million People Are Changing the World, Harmony Books, 2000 sociologist Paul H. Ray, psychologist Sherry Ruth Anderson

So don't despair, just let go of your fears and doubts, and show more of your full self, and do what you feel compelled to do. It is the road to joy, meeting many new soul friends along the way, while we create the best world we can.

The list below outlines key values guiding a *Cultural Creatives'* behaviour. Who do you recognise in here? You? Friends?

- Authenticity: actions consistent with words and beliefs
- Engaged action and whole systems learning: seeing the world as interwoven and connected
- Idealism and activism
- Globalism and ecology
- The growing cultural significance of women

Core *Cultural Creatives*, half of the total number, also value altruism, self-actualization, and spirituality.

Ways of thinking and values are key factors of *Cultural Creatives*, and the authors tell many stories of how groups have joined together, harnessing common ground, using communication skills and media, to bring forth ways of life more in line with their values.

Surveys were done in 15 EU countries in 1997 identified a similar number of *Cultural Creatives* in Europe. Current world guesstimates are at least 200 million, but so many countries have not been studied.

So if they're like you, watch out for other CCs and join creative forces with them. They tend to be involved in various groups and projects, often cross-pollinating them, and giving whole systems solutions to areas that have become limited by specialization. They consider both the traditional and modern ways, and the newest discoveries. Their resulting projects include justice, integrity, interconnections and long term sustainable answers.
This video[9] shows a taste of the possibilities.

---

[9] CULTURAL CREATIVES - THE (R)EVOLUTION for those who believe they are alone FogelMedia © 2016
https://vimeopro.com/kulturaliskreativok/cultural-creatives

Creating and blocking connection

Now we have some extraordinary insights and guides to use. But Life is still a mystery. And it is certainly a "Catch 22" for me to say that you will have to believe it before you can see it. Basically that is rule number one. And so it will mean that many people may be frustrated, and suspect that I am some kind of charlatan. But it remains, only you can prove these to yourself, by trying them out as well as you can, and seeing what works.

All of us come into some aspects of our lives as beginners, and want to jump immediately to being experienced. Not being yet able, we may persist or we may give up because we don't like feeling incapable, or we have a view that it's not worth the time to pursue.

But to gain these wisdoms, it's all about the doing. The embodied experience... that means the experience of consciously doing it, not just thinking about it.

Checklist for creating and blocking connection

1. What you put your attention on grows, because you are activating it.

2. Create connections in your mind, heart and emotions, and tell it as your story.... It will occur (It's not lying, it's creating. Because if you keep telling the old story, that's what will keep creating... e.g. "I want a man/woman in my life (but I don't have one now)", or "I want a job that I enjoy where I'm well-paid and valued (but I have one that I hate)!" The new story is more like, "I love my job, the people are creative and supportive, I feel valued, and the work is improving conditions in the world." (And yes, we will explore how to move to this unconflicted view.)

3. Find what is blocking you:

a. Beliefs that create counter thoughts. (see BRAIN WATCHING and BRAIN WASHING)

   b. Old habits of thought and action.... we are wired for habits, but we are able to change them if we're willing to add in new joyous patterns, and notice and release the old. (see FEELING YUCKY IS JUST A HABIT)

   c. Our inner hurt children, which will need extreme kindness and understanding, healing, and a firm hand. Some people refer to this as our Ego which likes to control. Others call it our "pain body" because the perceived pain level can be very high.

4. Situations in which we feel stuck.... we always have choices we can make to move towards our true goals of joy, freedom and using our gifts to benefit others.

5. Notice signs. (see READ THE SIGNS)

Of course these techniques can also be applied more broadly, though here we have been focusing on connection.

Our beginnings

When we were little we had times when we were lonely and perhaps felt like aliens, as though we didn't belong. Very few of us had experiences where we always felt part of the fabric of the universe. Did you always feel connected and as though "with friends"? If so, you would be rare.

Mostly there were adults around us doing their best, but we felt often alone and separate, and we had to learn to be strong, independent and often serious.

I certainly wasn't fed the lines that "life is a bed of roses", and "go with the flow" or "ask for the help you want and it will turn up". Nor do I recall hearing "decide on how you want your life to be, daydream yourself into it, and sure enough it will happen". It was more like "wake up and get real. We are all separate, and yes sometimes it's scary but somehow you will be alright."

Definitely there was a fear base of belief for me. Even though my parents were kind, hardworking and loving, they had these beliefs of course, and lots of emotions. They had lived through some major life challenges that were handed to them, including war. I am very appreciative of what they did for me, and I feel they did an amazing job.

It's just that this life has lots of denseness for everyone to deal with, which creates many challenges for us all as individuals and societies. Fortunately we now have the tools, and the understanding of how these tools work. Later in the book we'll look at working with energy and other tools.

Dissolving obstacles to connection

It is quite reasonable not to trust many things in this world. So let's consider the wisdom in what you can trust and what you can't. I was once told that you can trust everyone – as long as you know who they are. You can trust a thief. You can't trust a thief with your money, but you can trust that if you make your money accessible to a thief that they will usually act like a thief and take your money. This gives us more certainty if we know who we can trust for what purposes. Most people are combinations of many things, and are very complex. There are either:

1. those who feel connected to the web of life (or however they describe it)

2. those who don't feel connected

That one personal decision, often made unconsciously, determines many things for the person, including if they feel they are creators or they are powerless.

Do you know which choice you have made? Perhaps you switch between the two on some occasions?

You will remember in Chapter 2 where Zena, Isabelle and I are talking about our experiences, where some science experiments, explained so well in the book the Divine Matrix by Gregg Braden, show that:

1. We can affect each other, and send and receive emotions, instantly over long distances

2. Photons, the material that our universe is made of, is affected by the presence of our DNA which is our genetic material. (In a nutshell, humans affect our world, just by being.)

3. When our DNA is removed the effects on the photons remain

There are so many examples of this in nature, and in science. We can see the many, many ways that all life is interlinked, and we affect each other.

Now if this is true (and it is), we have never been alone. We've always been connected. We have been able to create, and receive what we created.

You really need to be willing to look at the evidence, and be open to seeing how it works around you. Because if you really decide that it is not true for you then, despite the overwhelming evidence, you won't feel or see the connections. What you believe becomes true... there is a later chapter devoted to this.

Yes, it's one of the practical science experiments that one must do for oneself. Many things can be read and heard and perhaps believed. But to be known, to be embodied and known, truly, one has to live it and experience it.

Create your own map, choose your own ending

This book is offering you insights into how this world seems to work. I don't want to create another system to try and see the world through, as there are already thousands of systems to map how life works. All of them are only maps. They are not the real thing.

I wish to point you to many existing systems that fit together, and simply lay out for you the main points that will probably support your own feelings. There will be stories to highlight how they can be applied. These are tools:
- To support your inner messages.
- To show you that the world has patterns and it is not just a meaningless mess.
- To show you many examples of how each of us sometimes work unintentionally against ourselves, and how we can work more in alignment with our intentions, and more successfully enjoy our lives and gain our goals.

There are an endless number of wonderful solutions in life. You can choose the ones that appeal particularly to you. The point is no matter how extraordinary each solution is, it is not the only answer. It is not the only possibility, nor the only way. There is always another way, even when one option seems so extraordinary.

So it is important for you to choose whatever feels good for you. That is the whole idea of this book: to give you general understandings of what works and then for you to choose specifically for you.

You may try many things, in fact you may put many approaches in place, or you may specialise... It is totally up to you as to what feels good.

In many ways this is like the "Choose your own ending books". Throughout this book you are given options as to which direction to go. And that is what life is really like, it's very much an interactive experience. We choose each moment how we are going to respond and we choose our intentions.

May you grow while enjoying your choices and your journey. (See Amy's story on working in hotels and how it transformed her life, in *First focus on what you want in your life*.)

As you will see in the next chapter, the choices you make become your truth. I don't mean this in a metaphoric way... they actually create your world. You'll see what I mean when you turn the page.

4

What you believe is true! yes, really.

We create our lives

As explained earlier, we create our world in ways that most of us would never have believed.

To do this we use:
- intention
- consciousness
- energy / vibration

These interact with all our parts including: our body; emotions; spirit; and our thoughts and beliefs (both the conscious and the unconscious aspects).

We are always creating.

We are always affecting ourselves and others, as well as our broader environments. We also allow ourselves to see and know only what we want to see.

So maybe not straight away, but eventually, what you believe becomes true for you.

Either your life changes, for better or for worse, or you deny the reality that is around you.

You can expand your life, or dumb yourself down, or do many other things. If you don't trust yourself you can make it so that others stop trusting you, and then you can blame them for their mistrust.... "Why don't they trust me?"

Yes, it's common for us to be unaware of our deeper levels and the effects they have on us and on others. Blame and mistrust are two of the results that we can greatly reduce when we choose to get to know ourselves better.

So let's look more into our great powers. We are all using all of these, but we're not necessarily conscious. So please read on and see what you recognise.

IMPORTANTLY: notice how powerful you are.

First focus on what you want in your life

"The more we push up against something, the more we find it wrong, and the more we wish it were different the more powerless we are to create the reality that we desire." - Karen Ann at Sacred Soul Healing

First focus on what you want in your life. Where your attention goes, energy flows.[10]

We often have the habit of telling the story that we are used to seeing as our life. Even if we'd prefer to live a very different life story, we want to tell our usual story, and in detail. We want to share that with others, and bond in our difficulties or pains of this human life.

Your life will improve if you CHANGE THAT.

This is why:

---

[10] http://www.noetic.org/blog/communications-team/consciousness-and-new

We draw to us what we are vibrating with. So whatever story we tell, "This always happens to me", creates more of the same.

How do we release these old stories and tell new ones? Ok, this is a challenge, because there are LOTS of ways. Also not so much because changing to new patterns is hard; it's more that it will be hard to persuade most of you that it's much easier and more accessible than you currently believe.

So throughout this book I am offering you a selection of ideas, and inviting you to choose your first method by intuition, and by being open to see if things present themselves to you as well. Synchronicities are those delicious coincidences where your dreams come to life or are propelled forward, with no effort from you.

You can create your reality... Here is one simple method.

1. Put yourself into a relaxed, easy state by getting your body comfortable (sit or lie down) and focus for 1-3 minutes on your easy and even breathing. (Stay aware of this while you continue on with this exercise).

2. Create a vision of something you choose to have in your future.

3. Make it as real to you as possible, as though it's already happening.

4. Spend time imagining you are already "inside it" living it.

5. Afterwards continue to create the highest vibrations you can. It's often best done by:

    a. Ignoring worries and doubts from your reptilian and limbic brains.

    b. Keeping distracted by focussing on other projects, passions, joys and appreciations, or giving to and assisting others.

6. Intend that the result will not occur in a particular way, but instead may well surprise you, and that it will be for the highest good of all concerned.

If you find that you're not doing this regularly, then enlist a buddy to join you. Organise a daily call for a while (phone, skype or whatever). Explain the process: you are to act as though what you want is already a part of your life, and you are daydreaming together as you are living it. Engage all of your senses.

After a few sessions it will begin to flow and feel easier for you. Let yourself feel how you want to feel. Feel your feelings, what do you choose: great happiness, or love, success? What do you see around you: a particular home; which people and how are they looking and acting?

Engage all your senses. Tactile: smooth, fresh sheets to lie between, touching your garden. Smell: how do you want your space to smell? Aromas of foods? Essential oils? Timber? Continue on with involving your other senses: taste, sight and hearing. If your sixth sense is real to you, what do your instincts tell you about this?

If this is hard, imagine you are in a parallel universe[11] where everything is as you choose - there are no constraints on you. Don't be too shocked, this is actually a viable possibility. Many scientists propose that we have parallel universes where alternate "yous" are living different versions of your life with others who are doing the same thing.

Set your phone alarm for 5 minutes and listen to your buddy's vision, become involved in their description where appropriate - it's like improvisational theatre, you say "yes" to all their ideas by accepting and building on them. You are helping them create this new part of their world. You may both feel shy at first, but live in your parallel world together for a while. You won't feel so silly when this new world starts to appear in your friend's life. In fact you both will be celebrating.

---

[11] 21 Ways to Get Unstuck: Keep Moving Towards your Dreams, Sharon McGann. Sharon and I collaborated on this exercise, which she explains in more detail.

Now set the timer for another five minutes, and now it's your time to have the pleasure of living in your new piece of world. Your buddy will support your creation.

At the end of the session make a time for tomorrow's session. Do this for a while until you are comfortable and easily doing this regularly by yourself.

This may seem so simple to you that it wouldn't be worth doing.

Nothing is further from the truth. Ask the people who seem most fulfilled in their lives and ask them how it happened. Many will tell you a tale of imagining it.

Amy told me her story of unintentionally visualising her future. As a teenager she was pleased to have work in a hotel, and meeting many travellers with interesting stories. But she didn't like the work or waiting on people, and began to imagine herself as one of the travellers, enjoying a life of travel and wonderful restaurant meals. She spent those years imagining her future life with passion, and it came about. Her studies and career took her in a new direction and now Amy is an international traveller, enjoying it as much as she had imagined.

We create our reality…. But what are the easiest ways?

Ok, we create our reality, but what are the easiest and most effective ways to do things?

Like a handful of yogis throughout history, we could choose to live without food, on breath alone. But only these few yogis have thorough belief that they can achieve this, and it has usually taken a lifetime to achieve. Most of us (and our predecessors, over millions of years) agree that certain foods contribute necessary factors to our bodies and health.

It is easier to go along with these than to change our deep-seated beliefs that we need to eat food. It takes a lot of attention and energy to change our paradigms (the set of beliefs through which we see and experience our world).

The questions then are: which shifts will be most valuable to you to make? And which changes are not worth the effort? That is a very personal question.

So how about you initially keep most of your life the same, but look at a few changes that will be important to take you into a new zone. I again emphasise the benefits of "adding in" gently before releasing any major life habits.

For example, if you don't have these happening, you may wish to focus on "adding-in" or improving some of these basics: nourishing foods; finding a renewing sleep pattern; ways to move your body for optimum comfort and function; more play and joy; and reducing poisons that go into your body. You will of course be looking to find what specifically suits you.

I encourage you to look at natural and drug-free approaches. But if your family has long relied on other methods, and is thriving on them, then that will probably be your path of least resistance. We humans can thrive on almost anything that we truly believe we can.

However, your intuition always trumps all.

Here are a few more clues about intuition. When you feel tingles, or great peace or joy, or other strong feelings, then you are receiving an answer of what's important for you.

Tim Minchin, comedy musician, and composer for Matilda the Musical, when writing his music and lyrics, searches for the tingles in his body. When he feels the tingles from that part of the song, he has succeeded in his goal. He knows from experience that when he has found something that affects him it will also transmit his feelings to his audience.

The way we feel our intuition varies from person to person. For many it's a "gut feeling". It may also be an awareness somewhere else in the body, such as sensations in the heart. Perhaps it comes as a conscious awareness.

Any of your six senses can be involved in delivering a message to you. You may hear, see, taste or smell something that's not actually there. I have heard sounds that aren't physically present to anyone else, especially music. And I have smelt things such as coffee while at work to alert me that I'd left the coffee boiling on the stove at home.

And remember the SIGNS in our bodies. These too are reminding us of our needs.

Thinking can block our access to our intuition, or at least confuse us. To bypass your mind, and go straight to your intuition, go back to the end of Chapter One, for the **Standing test to answer questions.**

Brain watching

Thoughts and feelings may pop up occasionally or often, to alert us to our deep down fears and beliefs.

They come from our subconscious which has been described as like the underwater part of an iceberg... 90% of it is not visible, even though that is by far the largest and most dangerous part because it is unchartered.

We tend to fear what is unknown. People often worry that their hidden depths will prove that they are "not okay" and "not good enough". However this is not truth.

Any so-called negatives that do appear, or are dredged up in personal development work, are signs of challenges that have been lived through. They aren't signs of a person being deficient in any way.

They do distract and interrupt what we intend to focus on. They are nebulous threats until the unknown becomes known.

One way to deal with these is to intentionally transform them. Write down any unproductive thoughts, feelings or beliefs that you catch yourself saying, thinking or worrying about. Julia Cameron in *The Artist's Way* calls them "blurts". Read on in Affirmations for the next step.

Brain washing

We are all brainwashed and all full of contradictions.

Wouldn't it be nice if we spent a bit of time, one way or the other, washing our own brains in the direction we want them to be, rather than just allowing The Winds of Life to blow through us? What if we actually took a hand in that?[12]"

"Wow that would take a lot of detergent." said Diane. "In many, many ways I have never been happier. This is like the best period of my life. Except for the fact that I am getting older and uglier by the day. But other than that I have the most skills in terms of social interaction, I am less up and down emotionally than I was in most of my decades - by far. And I live with my daughter and her family so I have certain limits: I just can't freak out, it wouldn't work. It wouldn't be acceptable with the children watching."

"But I feel so old. People don't experience me as old they keep saying. I don't know what that means."

---

[12] See Adding In and Fading Away

R: "That means that timeless part of you, your soul which has no age, is very strongly present. And you are allowing that to be seen. So you are not an old woman, even though you are over 70 years. You are a very joyous and powerful spirit. Obviously you are allowing it to be seen."

After a thoughtful pause Diane replies *"Yes I am at times. I am wondering if that youthful part of me that people see is all that authentic. Is it an act?"* With almost no pause she continues on, *"I think it is my child, the essential Diane. I think being a scanner[13] too. There is almost no human being that I am not interested in. Unless they are so armed that they can't play with me at all. Actually not even then!"* she has surprised herself, *"If they're not there with me I would be deeply interested in how that happened."*

It was time for me to clarify a core belief with Diane, so the conversation continued with me asking "I wonder if you believe that this is an intelligent, responsive Universe, and whatever we put out there and create, that is what comes to us? Whatever our energy is resonating with, from all our deep down conscious and unconscious thoughts and beliefs, whatever we are resonating with holistically, that is what we are creating in our life. Is that something that you could accept or not?"

Diane affirms *"I can. I can. I feel both ways about it and somehow I haven't overcome my own feelings that I have missed so many chances and I am a coward and I haven't overcome my own feelings. At the same time I see it working all the time. I see it working with other people, but I see it working all the time."*

R: Yes, we are speaking from personal experience and personal battles. So I really understand us thinking things like 'I should have written this book 20 years ago and I should not have missed so many opportunities'. It's what is called "shoulding" all over ourselves. But seriously, 20 years ago I didn't have the information; I hadn't worked out how to put it all together. I didn't have the big picture that I have now. I didn't have all the puzzle

---

[13] Scanner : a person who is interested in many areas. It's a term coined by Barbara Sher.

pieces. At that time I did the best I could do. I was raising a young family and dealing with myself and all the things that pop up that you have to deal with in your own self. If I'm really, really honest, instead of beating myself up about it, I have lived quite a fair life.

Visualising

Diane was taking in what I had just said. She was clearly open to hearing more, so I continued;

"So visioning techniques might work for you. Maybe writing stories about your healthy wonderful life, where you are ageless and have experienced everything you want to.

You know it's quite nice to imagine being 100 and writing a letter about the last 30 Years of your life and imagining it from a very positive View Point."

D: *"I think I'd go just a little more than ten years from now, I don't think I could imagine my life at 100."*

R: "Good, if you do what's comfortable for you, then you're not resisting yourself. You actually wire your brain as though you are living this future, healthy and energetic life.

And by not resisting you're not maintaining the old pathways where you dreaded ageing. These are fading away.

Exercise: reflecting on my fabulous life

So you write about the best life you can imagine, and what it has all been like. Here are some prompts to get you started. Write what you like... Dare to imagine!

1. What did you do and experience?

2. Share your feelings and sensations.

3. What did you achieve?

4. Tell about your family.

5. Tell about your important friendships.

6. What have you been doing with them all?

7. What adventures and holidays have you been on?"

Now for bonus points.... (Did that get your attention?)

The next step is to regularly recreate this future in your mind. There are various ways to do this, but the common first step is to get yourself into a good vibration. A simple method is to become mindfully present by being conscious of your breathing, or touching an object as if for the first time ever. Observe every detail and the sensations that arise as you touch it. Then choose from these:

1. Read your letter and experience it in your senses and awareness as though you are the older version of yourself, radiantly happy with your life. Do this regularly.

2. Record yourself reading the letter (I use my phone). Play it back regularly when in your relaxed or energised state. The goal each time is to feel how real it is.

Voila! You have done some wonderful brain washing. Now let's look at another way.

Affirmations

How do you feel about affirmations? Have you used them? Have they ever worked for you? Research in the area of brain plasticity has shown these to be very effective.

These are another way to refocus your attention, or to reprogram your brain and your consciousness.

Here are some that you may enjoy, favourites of mine and my friends:

"Where is the Magic?" "Let the Magic flow."

"How can it get any better than this?"

"I am truly divine, connected to all."

"My life is in divine order." "All is good"

"Let love and wisdom be united in me." (Catherine Ponder)

"I walk in paths of pleasantness, prosperity and peace." (Catherine Ponder)

"Show me the Magic!" "I am that."

"I can do it, I am a wonderful and powerful woman/man."

"If it's going to be, then it's up to me."

"Nothing stays the same, I welcome the gifts of change."

"What I make a shift towards, makes a shift towards me." (Robbie Zec)

"All is in Divine right order."

"I love and accept myself."

"Every day in every way I am enough."

"I am bountiful."

"Whatever I want emotionally from others I must be able to provide for myself."

"A journey of a thousand miles begins with a single step." (Lao-Tzu)

"I am my own best friend."

"Today is a gift - that's why they call it the present."

I enjoy turning my favourite affirmations into songs. This makes them easy to recall, and more likely to be repeated. Can you imagine how many hundreds or millions of times a negative thought flashes through our brains? So to balance that, how many times can we sing or say it back to ourselves? Remember that your energy and intention while doing it can greatly multiply the benefits.

One of the systems I like best to help with transforming our negative beliefs and self-talk is Julia Cameron's process in The Artist's Way. In her suggested daily practice of free writing, she encourages us to catch and record what she calls "blurts". Then to rewrite them into life-enhancing statements, and regularly read or re-write them.

You can take your unhelpful thoughts (did you catch some in Brain Watching?) and re-write them as affirmations.

For example, you might change "I am getting older and uglier by the day" to "I am an ageless and beautiful being, part of a divine universe" or "I am as beautiful as the trees and the stars"

Or change "I am too stupid to get this right" to "In truth I am connected to the universe and all its knowledge."

Affirmations are a clear form of brain washing. The important part is that WE choose our thoughts.

There are many other tools that we can use to change our thoughts and beliefs including: affirmations; Gestalt; Expressive

Arts Therapies including drama, narrative, music, art, dance and play therapies; and RET Rational Emotional Therapy.

What is most valuable for you to create?

Do you want more clarity on what creations are most valuable to you? To your soul? Explore these activities below, and remember this is your inner life, your gems, and keep it as private and sacred as you like. (Also see Playfulness and Words for Playful Exercise to remind you of your potential wide range of states of being. What an adventure!)

1. Top 10 Values - with the help of lists of Core Values that you can google, create your personal top 10 list of values for your life now.

    Prioritise them.

    Where does your life align with your values?

    Where are your life and values not in alignment?

2. Draw Your Dreams - no you don't have to be able to draw - stick figures and squiggles will convey your dreams just fine. Use colours... Let go! Have a go... What you learn will surprise you.

3. You're SO Talented! - Begin listing your talents, and at least once a week read it, appreciate yourself, and add anything else that occurs to you. They will cover the broad range of human existence, from cooking a great omelette to cello player, kick boxer, great listener, flowing dancer, warm-hearted, optometrist, parent, put people at ease, and great at puzzles. What talents do you overlook because they come easily to you and you assume that they are easy for everyone? Ask around and you'll find out that this is not so.

4. Write a Summary of your Life in your last vital years. Write it as a letter of appreciation (this exercise is more fully explained in Brain Washing).

5. Do one of these as a poem, or draw it, or act it out on a video just for you, or write a rap song (it's all about the rhythm, not the tune).... go on, try something new:

   a. Where do you belong?

   b. How is your self-worth? Self-talk? Self-esteem?

   c. What are your dreams?

   d. Write the life that calls to you.

Now if you have just read to here without time to stop and do one of these exercises, how about you pull out your calendar right now, and schedule a particular exercise in. Select a specific block of time and write the exercise number, e.g. "5a painting" and include this page number, so you can come back here at that time and be reminded of the details.

Doesn't it feel good to make a decision, especially one that allows you a creative treat? It's a special, exclusive time with yourself, and now you can enjoy looking forward to it.

We all begin life with our particular talents, and our own driving forces which dictate the meaning we are looking for, and our values. None of us can assess how another person lives.... only we know what is important and valuable for us. Often it's the apparent side-roads we take in life that give us as much, or more, than the main roads.

Just imagine what a world this can be.... "You may say that I'm a dreamer, but I'm not the only one. I hope someday that you'll join us, and the world will be one".... John Lennon.

I've been a dreamer for a long time. I know I'm joined by billions more.... what about you? And if each of us is doing what we love, and sharing that with others, then what a wonderful world we will build.

Trying to believe you are the writer of your life? Needing more proof?

Working with this book can be challenging because you may try something and it doesn't work. This kind of work takes persistence until the ah-ha moment occurs. To use a very simple analogy: it is totally dark until you stumble across the light switch, and then it is all bright and clear, and you now know how to stop the darkness any time you want.

This book is challenging some popular viewpoints which are now obsolete, but many people don't know this yet. There are many times in our history when popular beliefs have been found not to be true. It takes pioneers to make the new a part of our world. Join me in being one of those.

As I write this I am in an aeroplane 7,000 meters (21,000 feet) in the air, the clouds fluffy and white below as night closes in. Just over 100 years ago, only a blink of an eye in the history of humans, almost everyone believed that flying was impossible. No one believed in a future where it would be so common for thousands of planes to be in the air at one time carrying millions of people. This was as likely as magic flying carpets.

Let's go back another four hundred years to the 1490's when people expected Christopher Columbus and his crew to sail off the edge of a flat world. Yet his discoveries of "new lands" changed the direction of our world.

Technological changes in the last century have grown at an exponential rate, everything including: computers; travel; space travel and satellites; surveillance; communications; robotics; microsurgery; military; social media; entertainment; arts; data capture and analysis. Would these have occurred if people had listened to the doubters?

So please rebel against your own doubts. Rebel by applying the suggestions in this book. Show yourself what is possible. You are the only one who can.

Denial and doubts are a normal, survival-oriented protection - our reptilian brain (the first part of our brain that evolved) is programmed to stop us from trying out new things, to PROTECT US from possible DANGER. Newness is not welcome. But the familiar is accepted because it has already been shown to be safe. So you can see what your reptilian brain thinks of the ideas in this book. DANGER.... DANGER.... RUN AWAY!!!

So if that's what you're feeling, STAY WITH ME! I have ways to get you through this, and have so many good people to link you to. Right now make sure that you are breathing slowly.... and deeply.... as you read on.

If you are feeling like this is danger, it can show up in sly ways. You may just lose interest, or suddenly feel the need to clean the house, go to the pub, or many other things may suddenly become more attractive to you, or need to be done. You may say "This is all ridiculous!", and discredit the virtual mountains of independent and corroborating evidence from thousands of reputable organisations and individuals. You may find yourself feeling "It may work for some, but we can't all be good at everything, and I bet I can't be!" (This is another version of "I'm not good enough").

The truth is: it's you who needs to change. (It's always me, the others are just showing up to help me realise, or remember what I need to understand.) Probably you have unconscious blocks. At this point you're screaming for someone to prove to you that it's not a sham or a trick.

... So right now please go back and look at your answers to exercises in "You've Already Proven It To Yourself".

Are you still reading this? No, really, please go and look at what you've discovered. Because YOU are the only person you can really trust. YOU are the only person who knows you so well, and who can share your deep wisdom with you.

So please don't keep reading what I have to say until you've spent some time with your best ally, mentor and confidante, YOU.

While you go I'll sing to myself for a bit.

Oh you're back! Feeling better now? Have you remembered for a while WHO YOU REALLY ARE? Good. You have all the power.

Now let's continue on about tools you can use.

Happiness and Appreciation

You want to be happier? The research shows that you first decide to be happier. You begin to change to happier thoughts and ways of looking at life. And then, as planned, you actually become happier.

Research shows that people reach their goals because they first decide to become happy. Happiness is created by a decision or intention to be happy. It often appears that people are happy because of what they have achieved, but the achievements come as a result of the happiness intention.

Start here: Each day write 5 things you appreciate.

The benefits: This small action refocuses you on the best in your life, and puts your heart into its most loving, giving and receptive mode. This is one of your highest possible vibrations (there'll be more about this later).

If done just before bed, this is shown to reduce stress and improve your night's rest.

How often do you have to decide to be happy and what do you have to add to your life to change to a direction of happiness?

"On the topic of happiness," said Sue, "I have actually been pondering the "not good enough" phrase this afternoon, after having some ah-has this morning about how I make my goals unachievable."

"I understand the concept of deciding to be happier and.... I don't seem to know how to maintain that initial decision. So I think it is something more like a daily choice - today I decide or choose to be as happy as I can. It hasn't seemed to work for me as a one off."

"Good point Sue", I said, "Yes we keep setting our intention as often as we need until we've washed all of that old programming out of our brain."

"You have been programmed all your life, now is the best time to do your own re-programming."

As soon as you blame you give away your power

You give away your own power by blaming either someone else, or yourself!!

Think about that - it seems that one of those could be true, but not both.

So let me explain.

When I blame another, I am denying that I had any part in it. I say that "It's their fault. They have hurt me. They should pay"... Or something similar... I choose to see it as very black and white.

But the true story is more complicated. Usually at some level we are aware of the other person's limitations, or intentions, and we have chosen to participate with them. Often if we look back there were warning signs or feelings of what was to come. Perhaps those around us tried to warn us.

We also deny that we have attracted this into our life (by our denial, beliefs, thoughts, emotions and/or energies). We deny that we can change these things in our own life and create a preferred result.

By blaming another we deny our own awareness and intuition. We deny our potential. We deny WHO-WE-REALLY-ARE.

And that's only half of it!!

If instead you blame yourself, you also block your own power.

It might sound something like this.... "I am wrong! I SHOULD have done this better!" And usually it's followed by some kind of a tirade which includes "I'm so stupid! Why do I always do this? What's wrong with me?" And ending with a version of the total stopper such as "I never get it right. I will never be able to create what I want in life."

It's not true. It's not TRUTH. It is a habit of feeling yucky, and we can change that. (see Feeling Yucky is Just a Habit)

So next time you notice yourself blaming, check to see who are you being unfair to.

Can words deter you?

Now some people see the forces and connections as angels, some as God, some as gods, the web of life, and some as Mother Nature. There are many ways to perceive and believe your reality, and here are some more: fate; destiny; karma; your 8 bodies from physical to energy to etheric; an intelligent and responsive universe; humanists; Taoism; traditional Chinese medicine (TCM); Ayurveda; nature spirits or divas; Druids; pagans and many diverse spiritual and religious beliefs.

I don't plan to dispute any of these beliefs. It is not up to me to try to explain or decide the mysteries of the universe. Humans are designed to find meaning and we have many, many systems created which do just that. There are wonderful positives in all of those, and many points of agreement.

So please I ask you for your understanding and your assistance. My intention is to share some discoveries with you. New ideas that you may choose to add in to your belief system. So if I use a term that is outside of your belief system, I ask you to search for one of your acceptable words that you might substitute in.

For many years as an adult I couldn't quite come to read about "God". I associated many of the wars and unkindnesses in our past with religions claiming they were killing in the name of their God... often both sides of the war claiming this same thing. I knew there were positives too, but this was a sticking point for me.

Eventually I was encouraged by a friend to replace God with Good. And for me this was the best part of the spirit of life. That warmth of connection. For me it was a way of expressing my fledgling understandings of the web-of-life-that connects-all.

P.S. there are many ways to make meaning from what you read. One more view of the Christian Bible can be found in the writings of Catherine Ponder, Charles and Myrtle Fillmore and Florence Scovel Shinn who wrote books which took meanings from the bible using metaphysical interpretations. (Metaphysical Bible Dictionary)

I have no power against people who are "doing bad"

Q: I have no power against people who are "doing bad", so what can I do?

A: you can:

1. **Out-create them** - put your attention on what you choose, not on what they choose (see First focus on what you want in your life).

2. **Offer them better energy** (see We are all connected).

3. **Understand them** (see Nonviolent Communication).

4. **Live in your own world**.... think of it as an alternative universe (see Energy). There are high vibrations which do not attract lower vibrations. For example, when looking at emotions they can be placed on a scale (you can search it as "emotional scale"). Some highlights of the scale are unconditional love and joy are at the top, going down through interest, boredom, anger, fear, grief and apathy. It is a matter of how much energy /life force /vibration each of these states produces. The vibrations of them are as distinct as the frequencies for each radio station. With emotions, as with friends, like attracts like.

Of course we are so complex that you will always think of an exception to what I'm saying, but there are powerful truths here to consider.

I've certainly not mastered this, but I have been using energy awareness[14] for decades now. I have been protecting my family, and home and yard. I haven't had a burglary for many years, including for decades when we didn't lock the house. My car when on the road has been interfered with a few times, but not when it's on my property.

When living 'in our own world' the important thing is to accept all of ourselves, because whatever we are we will attract that to us... our shadow... fears... unconscious... layers which exist together but are contradictory.

We can choose to acknowledge and accept they are there, and learn to work with them.

---

[14] see The square root of 1% of people can create tipping points

What you believe is true so everybody is right!

What you believe is true so everybody is right! Haha - in an argument everybody is right.

As everybody is creating their own reality, argument is futile, so is sarcasm.

In her many books Catherine Ponder states the healing power of love, and countless authors have confirmed her conclusions.

Dr Masaru Emoto, with water and with rice, has shown that hate and love have different energies and vastly different effects on water. By the way, it's important to remember that humans are approximately 67% water. So how you speak to yourself affects your body and how you think and feel, just like it affects the rice and water.

In all my training with time, space and energy, and with other learning, it has been clear that the energy of love is a connector whereas hate builds walls and disconnection.

When you come from love and positivity it changes your world.

Each of us in our own lives have patterns, and it can be very easy to put our attention on worries and on what we hope won't happen. Again we forget that whatever we focus on grows. We can feel so stuck in these ways of doing things.

Until one day we become conscious of our patterns and realise how crazy they are, and that we have other choices. We realise we have more power than we had thought. This becomes a day for great celebrations!!

These are true for you!!

Here is a checklist which I recommend for you to read or recall regularly:

1. You are not to blame.

2. You are good enough.

3. You are complete.

4. You haven't wasted your time.

5. You have gained unique insights about life.

6. You are here, now, in the right place.

7. At any time you can decide to change your direction in life.

If you don't believe any of these to be true, then you are denying your true self and limiting what is possible.

You can choose to work on believing these keys about yourself. Look at all your counter-beliefs to decide if they are false, and if false then release and replace them.

As these keys become more true to you, your life will expand and evolve, benefiting you and many others.

To emphasise: these are true for all human beings of every size, age, colour, shape and ability.

All of these points ARE TRUE FOR YOU!! There are no exceptions.

5

Tune in and read the signs

Tuning In

INFORMATION IS COMING TO YOU IN VARIOUS WAYS AS YOU INTERACT WITH THE WORLD.

Believe it or not, you are a wise and wonderful spiritual being. You have many talents common to most of us.

You can move your awareness around, which suggests you are consciousness which isn't limited to your body.

You can be aware of how other people are feeling, or their energy, even when they are at a distance to you. This is usually greatest with those we're closest to, especially with partners and children.

But equally it is possible to "tune in" to strangers or acquaintances if you have a reason to do so. As a massage therapist I would do this before a new client arrived, and I would often sense where their pains and restrictions were. And while massaging clients I would at times feel their emotions and physical pains.

By "tuning in" it's possible to gain huge amounts of information as "first impressions" with people. And it's common to then ignore that, only to find out much later that we've chosen a path with them, even though we already knew there would be a big problem between us... Or maybe that's only me?

We can feel calmed or annoyed by each other's presence, words not needed. We can heal ourselves and others or add to our dis-ease. So it would be helpful to look at how signs can help us with all this.

Pain is a sign

Pain is not your natural state. It's showing something isn't working correctly.

It is clearly alerting you to a problem. It's like an alarm that immediate action should be taken to stop this problem or correct it.

Sometimes we are so scared of what the pain might mean that we avoid it. When pain comes to you, how do you deal with it?
- Do you cry?
- Hide away?
- Block it with drugs?
- Ignore it and "soldier on"?
- Do you worry?
- Or do you pay attention?
- Do you get professional assistance?
- Do you make changes and create a paradigm of a better future?

It is important to honour the pain, and this includes pains that aren't physical, such as emotional or soul pain. Often sitting with it, you will be able to discern something beneath the pain. Perhaps a feeling, or a memory, or a belief. Work with this underlying key, to work through it and release it. There are many kinds of therapists who can assist you with this, and many kinds of exercises you can do yourself to allow a deeper conversation with your inner wisdom.

Here are some possible causes of pain and methods to rebalance:

1. **Tight muscles from overuse or lack of care and maintenance**.... stretching, relaxing baths, massage, mental relaxation, yoga, tai chi, qi gung

2. **Tight muscles from imbalanced use of the body or structural misalignment**.... Chiropractor, osteopath, physiotherapist, remedial massage therapist, Feldenkrais, alexander technique

3. **Emotional pain.... pain stores in the cells**... we can work on the physical symptoms as in steps 1 or 2. Also address the causes: expressive therapies from art, writing, music and movement so you can become in touch with your emotions and your inner messages (see When you block your bad feelings..., and Checklist for vibrant, abundant, good emotional health). Also counselling, gestalt, nonviolent communication, creating the life you deeply desire, and energy work or spiritual development.

4. **Stress**.... This switches us to fight/flight etc, so that our energy is all directed to the priority of survival, and energy isn't available for our usual body maintenance and repair systems. If the stress is habitual then our bodies become under par... relaxation techniques, mindfulness and meditation to "be in the present", increase movement, hobbies, and add lots more satisfying and fun times.

5. **Spiritual pain**.... Our soul is trying to redirect us as we have not been paying attention to our deeper aspirations, abilities and passions. These pains may appear as mental dis-ease. They are often the difference between how we are living and who-we-really-are. Are we following our hearts? If you have no idea if you're following your deep urges, then it's time to use one of the techniques (in 3. In this list) to become better acquainted with your timeless and extraordinary self.

There are many great books written to help us identify the hundreds of signals that our bodies and minds give us. Each individual is so complex and multi-layered that I can only give you some broad ideas. You'll have to search further for your specific answers, and remember that your intuition, and perhaps also an experienced practitioner, will help you with your search to greater wellness.

Here are a few common situations with simple meanings, to give you an idea[15]:

Back pain	feeling unsupported or financial worries.
Diabetes	not enough sweetness in your life.
Eye problems	ask "what are you not wanting to see?"
Foot or leg	feeling blocked from moving forward.
Shoulder	carrying the weight of problems "on your shoulders", often those of others.
Sore throat	what aren't you saying?
Stiff neck	rigidity in thinking, or less mental flexibility.

Read more signs

We don't have to wait for pain or other messages that alert us that something is not going well. Here are some signs to notice:

1. If information or an awareness comes to you three times, pay attention. This can include:

    1. **An article or other piece of media** that randomly crosses your path, or that someone tells you about.

---

[15] The Secret Language of Your Body: The Essential Guide to Health and Wellness, Inna Segal

Heal Your Body A-Z: The Mental Causes for Physical Illness and the Way to Overcome Them, Louise Hay

2. **Words that you hear**, including in songs.

3. **Something that "pops into your mind", or an intuitive feeling**.

2. **A dream**, especially if it keeps repeating, or is highly emotional, then it's really trying to gain your attention.

3. Ask for **feedback** from our intelligent, responsive universe. Ask a question, be open to receiving an answer and watch and listen for it, perhaps it will arrive in one of the ways listed above.... Let me know how it goes.

While you are doing any of the above, if you observe yourself thinking negative thoughts about these signs, you are catching your negative self-talk. This comes from our usually unconscious beliefs. ....Quick, capture these words on paper (see Brain Watching) and you can begin to rewrite your inner habitual scripts to what you really want (see Brain Washing). And no, it's never too late, even if you are 102.... we all can change (see Feeling Yucky is Just a Habit).

Here is a message that most would miss

Your body is giving directions on how to be healthy and in balance. In fact nature is always giving directions too.

Here is a message that most of us would misinterpret.

Imagine I am mistreating my own body, and my poor body would feel so much better if I were taking better care of it. Perhaps I am overweight and tired.

When I look in the mirror I see the tiredness and weight and blame ageing. I feel stuck and blame myself for not having looked after myself better, and not knowing how to change. You could say I'm taking it as punishment for not being good enough. And I'm creating a downward spiral for myself.... This will not end well.

Let's take the same scenario and this time I receive the information, the same information, differently. I see a tired, overweight body as a cry for help. Each time I am aware of this I don't see a sad wrinkly future. Each time instead I see I need to be better cared for. I see that bringing myself back to a balanced and healthy future is within my reach.... I am making a start on an upward spiral for myself.

This is an opportunity to feed myself well, to treat myself with more kindness and perhaps more organisation. I could add-in some enjoyable changes. Perhaps it's time for some medical or nutritional assistance or something else. Perhaps I have been hiding the urgings of my soul with addictive behaviours including workaholism.

But basically it's simply a sign that I need to be taking better care of myself. And that will restore my health and inner sparkle, which allows others to see my natural beauty. It's as simple as that. There is no shame or blame.

Things that happen to us!!!!

How often do you ignore why things happen to you?

Our whole world is connected, and this is an "intelligent responsive universe", so responses from the universe are giving us messages.

What does it mean to be in a car accident? Break a foot? Strain a muscle? They mean we are being slowed down or stopped. The wise action is to look at your life and what is out of balance, or out of alignment with your values or your soul's urgings. Is there a part of your life that needs more attention?

I am aware of people with more than one major accident or illness.... if they didn't pay attention and see WHY they were being reminded to stop, then they may be stopped by something more definite telling them to pay attention to their life.

Okay, so now have I gone too far? How crazy to suggest that everything has meaning!

Most people agree that these are just bad luck. But I beg to differ. Consider this: why do some people seem to have a lot more "bad luck" of this kind? Sure, perhaps it's not them directly, maybe it's people around them who are very dramatic, or having quite a bit of chaos, or making decisions that aren't in harmony with how they feel or with the driving forces within them. Can you see that all these factors would be significant, could start a chain of actions and reactions that people will respond to?

Have you ever had the experience where you were ill or injured and actually appreciated having no choice but to rest? Perhaps with solitude? To be given respite from your responsibilities? The stress released? It can feel wonderful!!

I am not blaming anyone!!!

We are all on this journey of life. You could compare this to doing a course. Everyone in the course experiences it differently. Some begin fast, others slow, and that will continue to change throughout the course for some, or they will stay in similar positions. Some are stressed, some calm, some philosophical. Some question the process, and may rebel. Some trust, some feel capable, and some feel not good enough. And so on. Each of us is unique in our plans and approaches and values and priorities. The course is a learning experience all the way through. As is life.

The more information to help you, the better. Support with how to: study; get practical experience; plan and schedule your work load; brainstorm and write better essays.... all of this help will affect your journey in the course.

But it's okay however you choose to do it. Whether you get tutoring or use intuition, or go out to parties regularly, or get an outside job. Perhaps you discontinue the course because you've found something that appeals to you more. All of this is your choice. Everyone handles their life their own way. All choices are good.

However you choose to live is valid and good for you. However you live is your choice.

There is no blame. No shame. You are perfect as you are.

If you choose to use this study aid that I'm writing for you, so be it.

There are no requirements other than that you are alive.

This is an optional support program, which gives you some of the underlying games/understandings.

6

Caring for your physical self

Checklist: What does vibrant, abundant, good physical health look like?

Grade yourself on a sliding scale for each item on this checklist:

1. Feel refreshed when you wake.
2. Fall asleep easily and sleep well.
3. Clear headed all the time.
4. Don't sigh or grunt when sitting or doing "at home" tasks.
5. Supple, flexible body that bends easily.
6. No or few pains or headaches.
7. All senses working well and are pleasurable - touch, taste, smell, sight, hearing.
8. Have enough energy for your needs and sometimes feel as if you are joyously jumping out of your skin.
9. Feel the lack when you don't regularly use and move your body.
10. Feel sexual desire as appropriate, and enjoy love making.
11. Love/need spending time in nature.
12. Recognise that experiences, traumas and wisdoms are all stored in the cells and the consciousness that is your body.

To help, a physician needs to know what went wrong

Health is our natural tendency. Science calls it homeostasis, where all body systems are working to bring us back into balance.

It's important to mention here that sickness isn't sent by god. If anything it's caused by our disconnection from who we really are.... god or our soul and the web of life. It's disconnection internally and externally on any of our levels including our feelings and intuition.

When a physician diagnoses you, but can provide you with no reason why you have that illness or imbalance, then that is not a person who can help you. Run away fast, and find someone who can explain the problem and the solution. If how your imbalance occurred is a mystery to them, and there is no causation in their mind other than "bad luck", then there is no way they can help to bring your health back to it's natural balance. It is likely they intend to stop or reduce your symptoms if possible - but that approach can cause other problems. And that includes all kinds of health: physical; mental; emotional; energetic and spiritual. I have always felt this, and it's taken me on a good course.

It's tricky of course when we don't realise that we are dealing with imbalances. Our organisms work so well to maintain balance, that symptoms can often be masked. And we are all unique, so I didn't realise that my lower energy levels have been a lifelong indication of an underlying condition. Consequently it has caused some serious ramifications over decades. For one thing I wouldn't have felt such a need for sugar to boost my energy, and would have thicker bones if I'd known. But now I know and I'm feeling good, and I've overcome osteoporosis and lifelong asthma without medicines, with no symptoms of either now.

Eventually, at some point you cotton on that there's a problem, and go to the people who have shown they have answers, and get some informed help. It is of huge value.

The old saying "one sickness, long life, no sickness, short life".... meaning that often those who have a problem, look at their whole situation and take steps to keep balanced. Those who haven't needed to explore their health may get a nasty shock.

Grounding or Earthing

This is a whole movement. Google it. And this short clip gives[16] you a great overview of the benefits in a short time. It explains how so much dis-ease and pain in the body and mind can disappear by your daily contact with the earth. And if your climate makes that unappealing, there are sandals which will earth you, and even bedsheets which will do the same thing.

Walking on wet sand along the edge of the sea, or on wet or dewy grass has always attracted me, and I feel better. The dampness works extra well to assist electrostatic discharge from your body. With earthing, all contact with the earth has this affect.

Do you enjoy resting on the grass, perhaps looking at the clouds and what you can see in their shapes? Leaning against, or hugging a tree gives earthing benefits, as does reaching into the soil when gardening.

When I first started working in an office tower at 18, I would take my lunch onto the patch of grass out the front and loll on it as long as I could, unless I was going for a stroll. It kept me feeling more balanced and refreshed, and it wasn't a hard choice to make. It just felt good.

There are many documented benefits of earthing including using it to set or re-set our circadian rhythms to allow a good sleep cycle at night. And by the way darkness at night has shown up in testing to be more beneficial than having lights on, for renewal and healthy sleep cycles.

---

[16] How Quickly Does Grounding Affect Your Body? Laura Koniver, M.D. The Intuition Physician
https://www.youtube.com/watch?v=xZs0mnfTlZw

Each part of us affects every other part

How does sitting up relaxed and straight at the computer help my mind, body and spirit? And can it really?

Before I answer this please take a moment to notice what is happening right now in your body. Observe:
- relaxed or tight muscles
- tension
- pain
- cramps
- facial expression
- body position
- your breath

What have you discovered? Oops.... I've just checked me and yes I do have some over-tight, tender muscles, my jaw is a little clenched and I'm hunched over and a little lop-sided.

Ok, let's look further into this: unnecessary tension in your body is your enemy. The muscles are shortened and less able to operate. This blocks the free flow of blood, pinches nerves and reduces the messages they send, ultimately causing pain as a warning. So let's straighten up, breathe deep, release some tension and see what's happening now.

The blood, when travelling at its optimum is taking fresh oxygen and food to every cell in your body, from your fingers and toes to your spine, nerves, brain and eyes. It also removes waste from the cells and carries hormones which regulate all the endocrine systems in your body, from the pituitary to the reproductive, the thymus, adrenals, and more.

Now your lungs are free to fill with fresh air, unlike when your body is hunched over and keeping your lungs only part way open.

Now take another moment, and imagine you're breathing easily. Your body is at ease, and it's as though there's an invisible cord passing through each vertebrae of your spine. It continues up through the crown of your head and is hooked into the sky, holding you up with no effort from you, so your muscles relax even more.

Now that your body has relaxed, your mind can relax too. It's not possible to have a calm mind while your body is uptight.

Many of us do not even breathe well

Begin to notice, when you are in a stressful situation, if you are actually holding your breath instead of breathing. Notice also when your breathing is shallow, and is only in the top of your chest.

With a relaxed, full breath your whole belly moves. Would you like to try out this exercise today and see if it changes you?

1. Sit straight and take in a full breath into the belly.

2. As you breathe out relax your body, release the tension (you may be holding it in unexpected ways).

3. Continue observing your body as you breathe in your second breath fully and easily.

4. As you breathe out allow your body to feel lighter as you release stress and say to yourself 'I am letting go'.

5. Enjoy your third full and easy breath.

6. Go on with your day until you next remember and simply repeat these 3 easy breaths. Your breath is such a simple and powerful place to reconnect.

First it will reconnect you with your body, then your soul, then the rest of the universe. We learn the importance of Being. We learn that even the simplest thing contains Worlds within Worlds.

All of us can be explorers of the unknown even in a few spare minutes that we allow.

You need no particular place or equipment. Just be, observe and learn to unrestrict your breath.

Bonus home play: find a symbol for this breath exercise to come to mind. Draw it or find something to represent this and place where you will see it or touch it regularly during the day. Having a reminder to take your 3 easy breaths can make all the difference.

I just woke up like this

People often say "I don't know why I'm in pain. I just woke up like this". Sleep won't be the cause of their pain, though it may be the trigger. People come to me as a massage therapist and are amazed at the pain that they are suddenly feeling. I am amazed that they have no idea why. So let me explain.

Usually these same people take their cars to be maintained, but don't consider the importance of physically maintaining their bodies. Elite Athletes on the other hand realise it's important always to warm up and cool down muscles, and to stretch and be massaged. They rely on the health and well-being of their muscles and they must never let them get k n o t t y. It's vital that they must always be at their maximum flexibility.

When you suddenly wake up in pain, it's simply the last straw. For months or years your muscles have shortened and become knotted from over use. You've not had much of a maintenance plan in place of stretching and recovery.

You forget that things are tight, and get used to the dull aches and nagging pains. Then it's one day of misuse or moving the wrong way that brings it to the point of unbearable. Your body is now giving you a sign. It's clearly saying PAY ATTENTION! And it's so effective that you do pay attention. At least for a while. At least until it returns to a dull ache.

So whenever you wake up in pain it is a warning to you to start taking better care of your body, and especially your muscles and tendons. It's time to begin regular stretching, exercise, magnesium/Epsom salt baths, massage and so on.

This is simple and the improvement to your comfort and abilities may be huge. For stretching, explore yoga, tai chi, qigong, Pilates, body balance classes or YouTube clips. Buy a foam body roller for connective tissue release as described in *10 Reasons you Feel Old and Fat and How You Can Stay Young, Slim, and Happy*, Dr Frank Lipman, 2016

When we carry more weight than we prefer

Let's look at when we are carrying more weight than we'd like. First there are reasons to explore, then ways to deal with them. And right at the beginning it's helpful to sit and write, draw or imagine yourself and your life with you at your goal weight.

Here are two exercises to help you do this. This exercise is to explore the fullness of your views on your body (see Living in your desired life). This exercise is to train your mind to create your desired body

Here's the first exercise, if you enjoy expressive arts

You will use the format of the **Exercise - Your Young Adult Self** in Chapter 8.

Ignore steps 1 and 2, replace with:

Step 1. "Imagine yourself at your goal weight"

Step 2. Use as many senses as you can e.g. seeing, sensing, emotions, smells, touch.

Step 3. Write down all the words that occur to you, do not judge any, write down whatever appears."...

Steps 4-10 as written for **Exercise - Your Young Adult Self**

Here's the second exercise, it's visualisation/sensing

Do you know, and can you believe, that people have lost weight without changing what they eat and without dieting?!! It's true!! They have!! You can do it too. Intention, and imagining-it-as-done are your goals.

You'll probably have skipped over that last idea. But here's the deal I'm offering. You and I can both decide to start spending 5 minutes a day (more if you like) feeling how good we feel at our desired size.... Remeber this exercise shows you one way to do it: *Living in your desired life*. If you are hesitating, how about finding a friend to have a small adventure with, and try it out together... Have you given it a go?... Well done!

How did that feel?

Now back to my offer that we start visualising our preferred lives. They can be any facets of you, it's not restricted to your body shape. What do you wish to spend time programming yourself to become? .... So do you want to join me? .... OK, let's do it... And imagine you come back here in a few weeks, our conversation will probably go like this:

"Gee you're looking great!"

"Yes, thank you, and I have so much energy now! I'm so enjoying all the things I can do now that were too tiring for me before."

"Me too... I'm so enjoying swimming / my holiday / bush walking / feeling at my best.... "

It's also important to consider if any of these reasons may apply to you. Then use your intuition to decide where to start. However you move ahead with this, please remember my KISS approach - Kindness Is Small Steps.

Am I carrying more weight for any of these reasons?

1. Especially from emotions, fears e.g. of being hurt, if there has been abuse including sexual abuse, fear of letting people too close. **Protection from things in my life.**

2. **Dislike of my body**, up to loathing. A program of self-love is the solution, and finding and releasing all counterproductive deep beliefs (or anti-self, deep-seated missiles).

3. **There are toxins stored in my body,** and the fat is a protection for my organs. My body will cleverly resist fat loss until I cleanse and remove the toxins. (See more about this in the next section: Hope for serious illnesses.)

4. **Deep seated beliefs** e.g. "I am not good enough " is at the base of many, many issues, or its sister "If they find out who I really am, no one will love me, so I keep up a pretend exterior" or "I'm just not attractive/clever/capable/etc."

5. **Compulsive eating or drinking** instead of "seeing" and feeling my emotions. Shoving them down, silencing them. Hoping they will go away, but they never do until I look at them fully.

6. **Not "backing myself"** to live the life that calls me with passion... "Oh, I couldn't. I'm not (fill in the blank) enough."

7. **Not even looking up from my life** to ponder and discover what makes me happy, what makes my heart sing, or makes me smile. Am I doing what I feel others expect from me? Have I even checked in if they still feel this? Do I even know where all my inner policemen have come from? Are they really from me?

Hope for serious illnesses

For almost all serious debilitating or terminal illnesses there have been those who became well again.

Dr Joe Dispenza is one of many who have studied the common denominators of spontaneous healings. Of the 5 common denominators he found, one was the belief that return to health was truly possible, and another was a sense of connection or a certainty that there was a loving power which they could link to.

There are healers and practitioners of different kinds who have had higher than average success rates in treatments. There are enough case studies available to show that some people can "beat the odds" - and why can't it be you? This book is on the bigger picture of how extraordinary we are and what we are truly capable of. There is not room to link to all the portals for so-called extraordinary results.

I ask you instead to consider the scientific observations I have shared with you. Add to this what you already know of atoms being made of even tinier particles moving around (vibrating at certain frequencies) in a lot of space. All matter, including you and me, is made mostly of empty space. We appear to be solid, and we have come to believe this as much as we believe we are all separate, we are not made of energies and we don't have a lot of effect on our world.

The truth is that we are consciousness, operating an energy body, with the tiniest bit of actual solid matter. Moreover our intention and resulting vibrations create our reality.

So, wouldn't it be prudent to reconsider our beliefs on wellness and illness?

If serious illness touches our life, it can make us very afraid, and we don't know who or what to trust. I want to share some things with you that you may be able to trust.

I want you to know that statistically it's clear that with the War on Cancer, and the battle against diabetes, we aren't yet winning these fights. But I am so pleased to tell you that there are other approaches that ARE restoring people to good health. As we've shown, what we focus on grows. The War on Cancer focusses on cancer. A battle against anything is still focussing on it, and so it continues.

The new science takes a different approach to consistently gain a preferred result. I will briefly explain the ideas here, and if you're interested see my website for more information and to suggest specialists of many kinds who can support you on your journey to wellness.

The body ALWAYS does its best to compensate and protect us. When things are out of balance for us, and that includes our mental, emotional and soul health, then all our available resources are directed at returning us to balanced good health. Of course what our amazing organisms are working to return us to is abundant health, which includes peace of mind, and satisfaction in our life path.

You would be amazed to see how lovingly your body is working to stay alive and healthy. It is complex and very beautiful. And there are many specialists who focus to understand and work with this.

Let's look at people who you may consult to assist you. Top of my personal list are holistic GPs as well as specialised organisations which focus on the whole person including:

- Diet
- Lifestyle
- meditation or visualisation
- intention and beliefs
- Nature
- Movement
- Counselling and creative expression
- Particularly look at those employing the ideas covered in this book

So what is scaring you the most? For me it would be if no one could tell me why it is occurring. So I will endeavour to give you an overview. The best simple description I have heard of this is from Don Tolman. His two volume book: *Farmacist Desk Reference (FDR) Encyclopedia of Wholefood Medicine* is valuable. So let me give a simple and brief explanation of one common cause, perhaps giving you a new viewpoint.

First off, a day happens when your stomach/ insides don't feel good, and your body is trying to expel it. But you want to feel better so you take something to stop the vomiting. Body goes "Oh, we have a problem. There is toxic rubbish to remove but we've been stopped from expelling it through the mouth.... Ok now we will send it out though the anus". The diarrhoea also isn't popular, so another magic pill is taken to shut that down. "Ok, we can't get it out, that's not good! We'll have to resort to plan B. We'll create a rubbish dump area so that it doesn't pollute or damage any critical body areas. So put it over there in that fatty tissue." We end up with toxic dumps inside us. We can't lose weight because the body is holding onto that fat to hold the toxins in. The toxins must be removed first before losing weight is possible. So, dear one, the rubbish is kept in the body, and can cause problems.

It has been shown that an acidic body is one in which sickness breeds. So changing to foods that create an alkaline environment, may have a huge impact on you, but work with a relevant healer who has the knowledge and skill to monitor your progress. By the way an easy way to gauge your own acidity is to use pH test strips (from health food or gardening shops) with a touch of saliva or urine, and match the colour with the included chart for your result.

Believe it or not, your food is not as good for you as it was for your forebears a century or two ago. Sadly there are many reasons for this. But in a nutshell, we add a lot of poisons to the food production, and with GMOs (genetically modified plants) there can be pesticides inside the plant as well as other additions including DNA from animals. And many other poisons are also systemic, which means they go through the plant, not just on the outside where we can wash them off.

There is a relatively new harvesting method called "desiccation" where Monsanto's Roundup or glyphosate is sprayed on crops such as wheat before harvesting. The aim is to dry the crop out to make it easier and to increase the yield. Little of the spray would wash off before it reaches the consumer.

With meat, so many things are fed to them now which are unnatural and dangerous for us to eat.... anti-biotics, growth hormones, poisons, strange concoctions of other animals, etc.

Food is grown with super phosphates which are only a few basic minerals, with so many of our essential trace minerals missing. The farming system also tends to deplete the soils. And to top it off there are more and more dangerous chemicals leaching into the soils and entering our water system. I'm sure this isn't a surprise to you. Our planet is a closed system and many businesses choose to use many toxic substances. Even our homes put many down the drains, on our gardens and into the rubbish.

Much more than food

Eating is not our only area to address. There are things other than your food which will cause acidity in your body. These include beliefs, thoughts, habits and emotions, and these are caused for a variety of reasons. Imagine the effect on a person's body from regular rage, frustration, bitterness or hate - a very toxic or acidic brew - literally. Consider instead the positive forces of kindness and joy. It won't surprise you that the chemical reaction these create in your body is alkalising - putting your body in a desired state for good health.

But all can gently be worked with, and literally a happier body will result. Perfection is not our goal.

Fortunately our bodies can tolerate much less than perfection, and still do a wonderful job. You will be wondering where to start with all these. See Action checklist 1 for suggestions.

Start with the easiest first. Your body is in a crisis situation, so it will be very helpful to direct yourself back towards balance as soon as possible, without distressing yourself further.

Here are some places to start. And remember it's just for a while, it's not forever, because so many people have overcome so many serious illnesses, and you can be one of them.

Action Checklist 1: Assist in balancing your health

1. Watch movies or listen to talks or music that bring you in touch with your feelings, or make your heart lighter, or remind you who-you-really-are. Bonus points if they make you laugh.

2. Drink pure water every day. Add a water filter, especially one that removes fluoride.

3. Breathing/ relaxation/ meditation/ visualisation/ paint yourself well.... Find something that works for you.... explore YouTube clips.

4. Move your body some way every day... exercise, walking, stretching, yoga, dancing.

5. Monitor your alkalinity - buy pH testing strips at your health food shop and daily test your saliva or urine. Eat foods that help your body to become more alkaline http://www.acidalkalinediet.net/alkaline-foods.php.

6. Buy organic foods. If too expensive, try to do this for the "dirty dozen", or avoid them.

7. Wash all your fruit and vegetables, by soaking for 20 mins in water and apple cider vinegar. Then rinse. (if no apple cider vinegar, use whatever vinegar you have available).

8. Eat less – cut your food intake until you are experiencing hunger every day between your meals.

9. Visit naturopath or holistic GP to look for other imbalances or underlying causes. Ensure you have good levels of micronutrients: minerals and vitamins. Also ask for referral to a dietician if you need the support.

10. Have an Osteopath, Chiropractor or energy healer ensure your spine is not obstructed, and allows all the connections to each part of the body from the spinal cord to flow freely.

Action Checklist 2: Assist in balancing your health

When you are established with Checklist 1, consider this next, and remember to gently and lovingly add these as you feel ready:

1. Explore, read, research and seek help to look into possible emotional or spiritual causes of your dis-ease or imbalance.... Allow your intuition to guide you to relevant methods and practitioners of mind, body and soul.
2. Learn tools to live in your preferred future. Science has shown that we create our own lives, literally. What we focus on does actually grow and occur. Telling the old story will entrench it. Creating the new future will create that... but you may need help.
3. Find a group for support who have come through a similar journey to you.
4. Explore your passions and learn about or begin a new relevant project.
5. Begin accessing both your right brain and your left, and look at becoming more whole. See which you tend to be and balance out.
6. Connect with your inner child. What did you love? What do you love now? What brings you joy? ...These may be different to your logical answers. Sometimes we make a logical decision but we don't feel energised or happy about it.
7. Keep growing, learning, and sharing your gifts. These are the greatest joys, and inspirations to live a healthy life.
8. Begin to examine how you commit your precious time, and whether the life choices you are making are aligned to your values; if they aren't, what will you change?

Dirty Dozen and Clean fifteen foods

Here's the Dirty Dozen:

These have the most chemical residues on them, so if growing or buying organic foods, these are ones to focus on:

1. Strawberries
2. Apples
3. Nectarines
4. Peaches
5. Celery
6. Grapes
7. Cherries
8. Spinach
9. Tomatoes
10. Sweet bell peppers
11. Cherry tomatoes
12. Cucumbers

In other years beans, potatoes, lettuce, zucchini, cucumber, broccoli, carrots, kale, pears and blueberries have been included.

Remember you can reduce the poisons by soaking in water with apple cider vinegar for 20 minutes and then rinsing.

Here's the Clean Fifteen

These have the least chemical residue on them:

1. Avocados
2. Sweet corn
3. Pineapples
4. Cabbage
5. Sweet peas
6. Onions
7. Asparagus
8. Mangoes
9. Papayas
10. Kiwi
11. Eggplant
12. Honeydew Melon
13. Grapefruit
14. Cantaloupe
15. Cauliflower

In recent years watermelon, sweet potato and mushrooms have been included.

This information will vary by country and region, from year to year. Australia and U.S. use many of the same sprays as Canada. This list was prepared by Canada's Environmental Working Group in 2016.

7

Caring for the rest of yourself

Your checklist for vibrant, abundant, good energy health

Grade yourself on a sliding scale for each item on this checklist:

1. You know that we are all instantaneously connected by a grid or field.

2. You can choose the energetic effect you have on people and the world.

3. You are sensitive to your own and other's energy needs, and don't use more or less in communicating than is needed i.e. you don't overwhelm or back off from others.

4. You are able to notice your own energy, and also to tune in specifically to the state of your chakras (the 7 energy fields down your central head and torso) and to clear them.

5. You can easily adapt your energy, to be appropriate and sufficient for the task at hand.

6. You can expand or contract the size of your energy field (awareness) to the task at hand i.e. just to your desk area for study, and as large as the auditorium for a performance, and as large as the world for a world-healing meditation.

7. You energetically ground yourself each day, and set up energetic protection for yourself.

8. You notice when your energy is being affected by others. It can cause many distressing feelings and sensations.

9. You have learnt to remove other's energy from yours.

10. You ensure that no other person's energy is hooking into you, and possibly draining you, and that you don't unconsciously do that to another.

11. You are able to absently send love to any person/people/group.

12. The energy you give out to the world comes back to you, so it's valuable to you to be responsible for your own energy.

Energy

> *"Loving people live in a loving world.*
> *Hostile people live in a hostile world.*
> *Same world."*
> *-Dr Wayne Dyer*

Good Vibrations is not only a great Beach Boys song to dance to or cheer on the housework, good vibrations are actually key building blocks of health and happiness. So here are some tips on how to use them:

The universe is made up of vibrations and so are you.

You can also affect everything by learning to change your vibrations. For good or for bad.... meaning you can make your life worse, or change in ways that make your life more of what you prefer.

You can affect your DNA (epigenetics), your emotions, your immune system (neuroimmunology and psychoneuroimmunology), your health and vitality, bioenergetics, your state of mind (brain plasticity) and heart, your effectiveness. You can also affect your present and your future.

Believe it or not you can have some effect on those around you, and theoretically on every other living molecule on the planet... and yes, that does include the rocks, earth and water too. This is a participatory universe.

There are many ways to work with your vibrations, and a whole world to know, so here is some basic info to get you started:

You have energy centres in your body which you can learn to be aware of, and to help them to be working at their optimum. When they are not optimal they are GIVING YOU INFORMATION. (see Read the SIGNS)

INFORMATION IS COMING TO YOU IN VARIOUS WAYS AS YOU INTERACT WITH THE WORLD.

You can learn to consciously work with your chakras[17]: you can build your energy up, and can draw energy into your body, and into every cell, and even heal bones.

Here are some things to experiment with: yoga; tai chi; qi (chi) gung; energy healing; reiki; Kirlian photography and spiritual healing.

This is an important area. Our energy centres can leak energy. We can be under par, with less energy. We may block our various centres from working freely, for all kinds of reasons, and this will have impacts on our health as well as our effectiveness. If, for instance, our third eye energy centre is closed, then we will not be aware of what is really happening - we will be "blind" to the truth, and more likely to either feel we're a victim, or to blame others. If our heart chakra is blocked it can diminish the life force and connection to love, and it can reduce our ability to bond with others. The heart has many other roles covered elsewhere in this book.

Of course you can live a long and happy life without being conscious of your chakras, but most well and balanced people will give their chakras a good work out through various ways, though they may not focus on the energy aspect of what they are doing.

---

[17] Barbara Brennan, *Hands of Light*, practitioners who have trained with her are all over the world and help and train people to work with their own energy and energy centres.

Here is a brief introduction to your chakras:

Root chakra: red, safety, connection to the earth

Sacral chakra: orange, creativity, children, sexuality, partners, creative projects

Solar plexus chakra: yellow, will, empowered action, making dreams come true

Heart chakra: pink or green, love, connection to others, friends, sharing, community, compassion

Throat chakra: sky blue, communication, speaking, manifesting through sound: talking, singing, conversation

Third eye chakra: indigo/dark blue, seeing the truth, the truth beyond the physical

Crown chakra: violet, gold or white, connection to the universe and compassion

Here is an example of someone who may benefit greatly from learning about chakras & energy:

Jules runs performance classes for adults, and becomes very involved with supporting their successful development. In our conversation I identified that in her workshops she was taking on their energy and feeling drained.

R: "Have you worked with energy, chakras and protecting your energy? For sensitive people, that can drain your energy so much. I've had energy leaking out my chakras and I've taken on other people's energy which is exhausting."

J: "That's how I feel... I feel like I've taken on all their gunk... and it's all over me."

R: "Do you notice when it's happening, or just when your class is finished?"

J: *"Yes I notice when it's happening because I get angry."*

R: "Good that you notice. Now you just need techniques so you can put them in place."

But Jules was so relieved to be able to talk about this very real phenomenon in her life, and she continued explaining. *"And I get frustrated! My body tenses right up. And I feel like I want to hit.... Of course I find an excuse to quickly leave the room and calm down.... Afterwards I am so exhausted, sometimes I can't do much more than rest for the next couple of days."*

R: "You're getting a very clear message. It's saying "Jules what are you doing? Why aren't you protecting me from all this gunk? I am so angry that I have to deal with this! Please stop it so that I can do more of the work that I love."

We all have energy, and we all can learn to work with it so that our life is easier and better. My mentor has a gift, which has taken her most of her adult life to come to terms with and train to use. She is very sensitive and can see, feel and work with energy, as well as receive knowledge for people, from another dimension.... she just knows.

In the classes she runs, the first thing she does is get participants to check into their energy centres or chakras. We tune into them, learn to know what is happening, and get them working optimally again. Often we discover life patterns that aren't helping us, changes to make in our beliefs and habits. We also practise not taking on the gunk of other people's energies. I've been doing this for some years, and my energy and life are much clearer and more enjoyable than previously.

Our energy and emotions affect each other

"Emotion affects your actions, thoughts, and connectedness, and can pull you out of the present moment. It stores in your body cells. If not explored, your unconscious can shock you with an emotional explosion from your amygdala."

Ben was following me until I said "And these often haven't even come from you!! Imagine that." The look on his face made me add, "Okay, now you think I'm making this up? Please stay with me a bit longer..." But I could see he was amused, and he burst out with *"Well how would they just get to me then? Sneak onto me in the middle of the night? Like a cat burglar?"* He was joking, but crazily he was right!

"Yes", I said as I laughed. "Yes, that's right... Or they might accost you in a busy supermarket. Or even hug you and put sticky goo all over you." Ben was laughing as he imagined how unlikely these were. "Does this sound ridiculous?" I asked, and continued before he could do more than nod, "But seriously, all these things and more happen to you. But let's break it down to show you."

You have hormones and other chemicals running through your body, and just as electrical wiring in your house creates an energy field[18] around it, each person has a field around them. Trust me for now if you can, until you explore more. (EMF: electromagnetic field).

Think of emotion (e-motion) as energy in motion. Awareness of how to deal with either one (energy or emotion) will greatly help you deal with the other. Our energy field content and our emotions fuel each other, and also whatever we put our attention on grows, so if we're not aware they can sometimes grow like wildfire.

So there we are, having all this energy, but not knowing. Imagine a person who hadn't washed or groomed their body, teeth

---

[18] Nishant Sharma - Co Founder BioField Global Research

and hair all their life.... would we be keen to sit near them? Not @!!#?! likely!!!

So if you have this energy that you haven't noticed, and it may be the equivalent of being unwashed and ungroomed, then it will be having all kinds of effects on yourself as well as on other people! And theirs on you!!

Energy is real. It can be measured. Our thoughts and feelings and beliefs create energy. Others can feel it.

We can feel theirs, but often we think all that is ours too.

It's time we learn to notice our energy. And notice what's not ours.

And better still to be able to keep ourselves clear of carrying around others' junk energy.

It will feel fresher, simpler, and so much tiredness, yuckiness and unwanted feelings will disappear. Note that grounding/earthing has a great effect on clearing unwanted energy, and many of your actions or activities will also clean up your energy. The biggest challenge is when unknown things are having very real and unwanted effects on our lives.

10yo boy needing energy protection

Recently I was speaking to Kevin, a 10 year old friend, and this was his concern. *"Sometimes I just feel yucky and I don't know why. It happens suddenly, usually at school."* So we had a chat, he answered a few of my questions, and this, basically, was my response.

"This sounds like other people's energy is affecting you. It is very common. Everyone has an energy field and it's full of how they are feeling at the time. You know how it's so good to be around some people? It feels so welcoming. Some people might make you feel happy being near them, with others you may notice you are quieter, and maybe feel tension in your jaw or head, and maybe your gut feels bad. These people aren't intentionally spreading bad feelings, but you sound sensitive to them.

So this is what to do. Every day when you have a bath or shower, imagine any yuckiness in you as a dark liquid flowing out and into the water.

Then afterwards imagine you put an invisible shield around you of white light or love. It is your bubble. Push it out to 2 feet above, below and all the way around you. It lets in all the good feelings, but filters out all the yuckies.

This is how many adults manage this. It's important, and it's great that you are already sorting this out at your age. Most of the yuckies you'll feel in your life do come from others, and now you're learning you won't have to accept any of that.

Good on you! Life will feel even better now."

And yes, life is supposed to feel good most of the time. And even great!!

With change it is important to be moving towards something

With change it is important to be moving towards something, much more than focussing on what you're moving away from. We humans find this more fun, more inspiring and easier. It is our natural state.

And what you put your attention on grows, so you need to focus on where you are going to.

Thinking of doing something is as real to your body and brain as actually doing it. There have been numerous studies on this, including imaginary piano and basketball practice. In both cases the imaginary practice gave equivalent learning results to the actual physical practice.

I have imagined giving myself doses of flower essences. This is a great example because they are a vibrational remedy. This imagined dosing has been muscle tested (using kinesiology - the muscle will be strong for a positive result and weak for a negative) and found to be effective.

Whose thoughts am I thinking?

How many of our thoughts, beliefs and impulses are our own?

Let's notice some of the influences on us. Please consider these next questions before you read on.

How many advertisements have you seen in your life? Yes, that is too hard, so let's break that down. How many would you see in a day or week? They are all through print, digital media and signage. And that's not all.... what reading do you do? What is the bias in the media? In the journals or books or YouTube clips, shows, movies or music? Most days there would be hundreds of messages that pass by our eyes. They register in our unconscious, even when we don't recall them.

So how many days have you lived? Let's approximate:

1000 days for every 3 years of your life. And we'll be extremely conservative and say 200 pieces of biased ideas or images reach your subconscious each day via your sight and hearing. (The actual number is much higher.)

If you are 21, that's 7x 1000 days = 7000 days,

And 7000 x 200 = 1,400,000 pieces of other people's views. One and a half million! At least.

So again I say, whose thoughts are you thinking?

Here's a game when you have time to watch yourself, or play it out loud with a few friends. You say something, or watch your own thought, and ask "Where did that idea come from?" Wait a moment and an answer may occur to you.

There are so many ways to look at things. Always we are affected by how we perceive things: our politics; social status; family; background; education and so on. As far as intentionally widening your understandings, and feeding your own interests, please consider:

1. Have you sat down and had discussions with others about the views and values of philosophers, activists, indigenous people, educators, economists, healers, social law and equity activists or professors?

2. Have you travelled and lived with, or gotten to know people with very different lives and values from you?

3. Are you discerning about the source of your media "diet"?

4. Do you specifically choose to have input from people you admire and wish to learn from or be inspired by?

5. Or do you take your information from whoever is there, without thinking about it?

The diversity and satisfaction of our mental diet has a huge impact on our mental health and our overall wellbeing.

Your checklist for vibrant, abundant, good mental health

Grade yourself on a sliding scale for each item on this checklist;

1. Able to focus your attention on things that you choose.//
2. Can settle intruding thoughts.
3. Able to control depression, anxiety and panic attacks.
4. Aware that you are the creator of your life, and you can widen your options at any time.
5. Acceptance of ourselves and others allows us to let go of shame, blame and guilt.
6. Can recognise your feelings and name the needs that underlie them.
7. Focussing on positive thoughts, and on what you wish to create, because you know that what you put your attention on grows.
8. Consciously notice your words and thoughts that show you have counter-beliefs, and are willing to release them lovingly while replacing with your own positive affirmations, visualisation, etc.
9. Being flexible of mind.
10. Enjoying a sense of fun.
11. Realising that serious jobs and topics can all be approached in a fun or light-hearted way.
12. Feeling "good enough" (often).
13. Feeling hopeful and optimistic (often).

Boundaries

"To the best of my understanding, life is a heroic attempt to find our innate strengths and beauties, even though we have been taught in so many ways to deny them." - Regina

Setting boundaries helps us while it also helps those around us. So don't be shy staking out your claim. Though I recommend to do it as kindly as you can.

Have you ever received a message in person or digitally, and felt bad? You were tempted to respond and make it clear they'd over stepped your boundary? You hesitated.... Then if you replied, as clearly and lovingly as you could, did that yucky feeling in your gut or heart change? I did it today and immediately my gut felt fine again. I was also pleased that I'd supported myself.

It may be as simple as reminding someone of the power of kindness or manners. Or letting them know that you weren't comfortable with their communication.

Our cell walls or membranes can become very weak, a physical manifestation of our lack of strong clear boundaries.

Similarly if we are not standing up for ourselves we can have problems with our spines, and/or our bones can become weak (e.g. osteoporosis).

We certainly have the power to create our own self as the window to our soul.

Can I show Who I Really Am?

Can I show Who I Really Am? Or another way of saying it is "Can I present my authentic self?" We'll be looking at this, and answering questions including: Which situations? When do I and when don't I? Why? What is best for me?

There is room for all of us to increase how much of our true responses we let loose in the world. We wear some masks for various reasons including protection. These are virtual masks that we hide ourselves behind.

Benefits

First let's look at some of the benefits of showing *who I really am*. I'll start with the story of a client of mine who we'll call Lucy. She is a vegan and works for animals to be treated with love and respect. She is and has been a long term yoga student for her well-being and her spiritual growth. She's also highly trained and skilled in business.

But a few years ago after thinking about it long and hard, she decided to retrain in something that she values and loves, and clearly is naturally suited to. She gained high quality training in counselling, both in theory and practice. She created a website and brochures with great care to describe her services, and our session was soon after her new business was launched.

At that point she felt surprisingly flat.

After I questioned her, she realised that she actually felt imprisoned by this business that she had created.

What she discovered was that whilst she was doing her counselling training she'd taken on the belief that it was very important for her to be a clean slate with her clients so as not to influence her clients by putting anything of herself into their situation. This is the ethical way to counsel and mediate. However she'd taken it to the point that she also did that in all her promotions about herself and her work.

So peace, her yogic spirituality, and everything about her core values.... none of that appeared in her promotional material.

Lucy had felt imprisoned because she was attracting clients into her business who didn't resonate with conversations about spirit and being the best people they could be. And the thought of dealing with people who didn't share those beliefs wasn't fulfilling her soul purpose.

She was so shocked that after all the growth work she'd done in her life, she ended up doing something so against what she meant to do. There was a lot of shock coming off in the session but at the end she came to that point of understanding.

By the end of the session there was a real transformation. Energy had lifted and she was really excited about the possibilities raised. Mind you she was also venting her frustration of the work that she would have to do in rewriting her website and brochures!

When we really know who we are, and what we value, we can put that out into the world and attract people who resonate with us, personally and in business.

Going further in that vein I have had some great help from Tad Hargrave, a Canadian with a business called Marketing for Hippies. Tad gives a lot of information about niche marketing which isn't pushy. It is based on being vulnerable; showing who you are and what's important to you. If you open yourself up people will think, "He's like that, I'm like that too. He's the man for me."

All the work that you do together is going to be so much more effective and joyful.

Another real benefit of authenticity shows up in Nonviolent Communication. At the core of it is being vulnerable to show who you are and what your needs are, to be able to come up with alternate solutions.

When showing who you really are, that is showing your integrated self. We are people with integrity, because we show ourselves, and that includes our shadow side which is a whole topic in itself.

Integrating my shadow

The shadow side I define loosely as the parts of ourselves that we think are not okay. Maybe it's the weight that I'm carrying and I'm embarrassed and uncomfortable about that. Or maybe I get cranky at people and I don't want to ever tell people that I'm like that. But integrating and accepting those parts of ourselves, and allowing with discretion to share this information, can make for more whole relationships with people.

Discretion is an important word here.... to share appropriately with those who are a safe place, ready to hear it and where it will make a difference.

Here are a couple of quick stories about my shadow side. When I was first married in my twenties it was as though somebody slipped a switch without telling me, and I found myself acting in ways that I had never expected or planned. I felt that I had to do things this way or that way because now I was a wife. And I share this with much emotion because it was a shock.

Another thing that happened to me unexpectedly was around that time I used to perform, and whenever I was given an authority role my body would become as stiff as a board and I would become like a caricature. I could be fairly realistic in other roles but as soon as it was an authority figure they weren't real people to me anymore, they were almost cardboard cut-outs that I was acting.

So it's really interesting to find out that there are things deeper within us that are coming up instead of what we have expected.

If I can own all these parts of me then I can work with them and move forward. That's really important. What I'd love to do is ask you if these stories and thoughts have prompted any times for you when you feel you haven't shown your authentic self, or you have seen that in others.

To define the authentic self, it's the integrated me, it's who I really am, the wholeness of me. I am willing to show more of me than society often seems willing to accept. In fact Brene Brown, who is currently very popular, is saying that real bravery is to be vulnerable. That's because being vulnerable is scary. Really showing who we are in public, so that people can possibly reject us.... that brings up great fears for many of us.

If we don't show our true selves then nobody could actually reject us. We can justify it and explain it away and feel protected, because however they responded it wasn't really to us, so we haven't really been judged. That's how many of us work it out often unconsciously. But equally by doing this we don't get any of the amazing benefits of truly being seen.

If you felt loved in your early years, you may like to remember back briefly to a time when you were between 2 and 5 years old. To recall how it felt before the time that you went to school.

I love that if you go to most kindergartens or preschools and you ask the children "Do you dance?" most of them will say "yes". "And do you paint and draw?" "Yes" they'll say. "Would you like to sing now?" "Oh yes, let's sing". There's no censoring of that creative side at all.

And then you speak to a teenager, let's say at 15 years old, when they're so very self-conscious and aware of fitting in. You ask them the same questions and most of those answers will be "no" won't they? Sadly it stays the same for many adults through much of their lives. So I am very interested in helping us to get back to the willingness to express ourselves we had as young children.

Do you have any stories of where you don't feel safe or you're just not comfortable or for whatever reason you have a particular persona to get the job done?

Feeling safe to be fully me

Kelly: *"I was very lucky I worked in a place for 8 years where I could be my authentic self. It was a place where we were free to let people know what we thought and be ourselves without fear of judgement. When people said things others just assumed that people were saying it to try and understand things or to make things better.*

And then I went to work for a place and I don't think anybody there was authentic. At first I didn't realise, and I couldn't quite figure out what was wrong. People didn't laugh. People didn't dance or sing. There were never arguments or disagreements at all. But it was all in a very passive aggressive state.

After several months I started realising. I was sitting in meetings where there would just be a lot of Yes-Man kind of things where the answers would always be "Yes, yes, yes". But nothing would really get done because no-one was really discussing issues and nobody was saying that they weren't buying into it. It was so strange, strange...

It was very weird for me. For example it was an environmental organisation yet most of the things they were selling in their stores they bought from China, and they were buying the cheapest products. And so I asked things like "What do we know about these organisations where we are getting the rubber balls produced for instance?" Or "What kind of chemicals are in them?" I really felt I was on my own when I brought up these kinds of questions. Responses were like "Why are you asking those questions? You're not in charge of the store." It was so strange: people were being so inauthentic, trying to fit into this organisation."

R: "That is a good example. And how wonderful to have that previous environment for 8 years. What a wonderful grounding."

Kelly: *"This was so dysfunctional though, and there was such great opportunities for that organisation and so it was so hard for me. How could the senior executives say one thing and do something completely different? How can people spend their whole lives in organisations like that? How can you hide yourself your entire career?"*

Kelly is just aghast as she ponders this situation.

R: *"Yes, looking at that first business, we can appreciate all those people agreeing to work transparently and playfully. Then you look at all the times and the places in our world where many government and profit-making organisations are playing similar games."*

Diane: *"I think that hiding my vulnerability is pretty much a full time job with me. It has become almost my person rather than my persona. But when I have tried to work in organisations in the workplace I don't really fit. It's hell for me. I can't even fake it there.*

I do teach at a hospital now, and that is plenty of strain. I was thinking as you were telling stories and speaking, when I teach I feel I am performing. It feels pretty authentic though, because it's so impassioned and well-informed. But whenever I get negative evaluations such as "I wish she was more science based", that's because they don't like the science that I brought. They want other evidence, they don't like the evidence I bring.

In the hospital I have to mind my p's and q's. When I am with clients in their homes I tend to be very familiar and overly unguarded, depending on the setting. But I'd say that generally I walk on eggshells in life a lot.

One of the greatest experiences that I have ever had is those professional practice groups where people really get real. It's so powerful. I'm not in one now but 11 years ago I had been in a practice group for several years.

I've been pretty much in hiding, various degrees of hiding all the time. But I've structured my life, at this stage of my life, where I don't need to hide too much."

R: "So you have found environments and ways of doing things where you are free to be you?"

Diane: *"Yes that is really very true."*

R: "It's because those other environments don't suit your needs. When you tried to be you in those other environments it didn't work, it just wasn't a good match."

Diane: *"I know, yes."*

R: "I would say that would be true for everybody. There would be some places that feel good, where you are very happy to open yourself up, and some places where it's not. For instance the second place where Kelly was working, you would have to classify that as a toxic environment. Seriously. And opening up in a toxic place is counterproductive. It is not going to help you. It is really not going to be a benefit. You are just going to get slaughtered one way or the other.

Unless we are incredibly strong, then we all need to find environments that are appropriate. Somebody once said to me "Don't just give out your gems and don't just hold back" i.e. don't do any of these automatically but discern where are the safe places to be who you really are…. Make sure that if you have gems to share, you share them with people who are worthy of receiving them."

Diane: *"I remember working hard at certain places because I love the work. The other workers would tell me don't do that you're making me look bad. You have to slow down. Again that's a tricky one: I was just working with enthusiasm and vigour, when everyone else was in the program of the institution."*

R: "They weren't comfortable with you being your full self, because it made them look bad. Our society does tend to be uncomfortable with differences, and this is why it can be such a relief to find people you fit in with, so you express your whole self.

Now, I'd like to explore those times when we feel very safe to be who we are. Those really good moments, those outstanding times. It was suggested to me that even within families there can be questions like "Do I wait till the kids are out, or do I wait till my partner is away, until I turn the music on and dance freely?" Is that the only time that I do that? Or when do I just let my heart make the sound that people may or may not call music? When do I sing my full spirit? Do I only ever do it alone? Do I do it in front of other people?

I usually just sing by myself, maybe when my partner is around, especially when I'm in the shower or bath, my favourites. It's a stress relief for me. The other day I was with 3 friends at a small house-warming for a friend making a new beginning in her life. I knew and trusted the three of them very much. My friend asked me if I'd like to sing to them. Nobody ever asks me that, but she did and I agreed. I went very quiet and allowed the music to come out of me without any plan.

That was a lovely thing to do but it has taken me decades to be at that point, and it is only something that I would do when I felt very safe and loved."

Diane: *"Some people I can truly connect with, but others can totally ridicule and mock me. But I believe the world is in dire straits, so people standing up with their truth and their parts of the solution is a good plan. Now it sounds to me Kelly that you have come from an area with wonderful opportunities to live your life openly and fully. So you would be coming at this from a very different angle from where I am and have come from."*

Kelly: *"Yes it's very interesting for me. As a child I was always encouraged to do whatever I wanted to do. I had parents who never made me follow through on anything. If I wanted to take tap dancing I could. If I wanted to go horse riding I could. And my mum used to say you're only interested until you get the uniform on and the equipment. Laugh. Then I lost all interest. But they still endorsed *it you know time and time again one interest after the next. But I was allowed to have that freedom.*

So for me now my current business is a huge challenge for me because I want to commit to this and I've got off to a great start. I am just the kind of person who could after all this work not follow through. I don't even know if this is a bit against my authentic self, but I am going to follow through this time and get this work done.

Also about authenticity I thought about when you said when you were first married. In my current relationship, I am definitely a lot sillier, definitely more my child like self. You know, I go jumping on the bed now and again or I will do a twirl.

Right now my guy is watching basketball which he's recently started watching. I always make fun of him because for me once you've seen it you never have to see it again. Nothing happens except people put a ball in the hole and one team gets a bigger number than the other. I do not get why you would watch that again and again and again. (She laughs) So I will get in front of the TV and commentate and just be silly. I would never have been that playful and silly with my first husband. (More laughter)"

R: "Does he still laugh even in the middle of a basketball game?"

Kelly: "*Yes he does. He copies me sometimes too.*"

Competition… to be like someone else

Diane: "*On the subject of competitiveness… as the younger of two girls my parents were always comparing us. "Why can't you be… this, this, this?" "Why can't you be more aggressive?" etc. I hate competition. I love the Olympics, but I think its horrendous how everything has to be scored. It's all about who is better. For me it's both boring and nauseating. It seems to me that Western civilization has only a few plots. The main plot is betrayal or infidelity and the other part of course crime and then competition. It's exhausting and discouraging. And it's why 10 year old children won't sing I think. I did read a study a few years ago that said children by the age of 7 were saying 'I'm not good at that'. Wow!"*

R: "That's another whole topic: education. I home educated my children for 3 years. With schools, because they measure people, and because kids are put ahead and behind and all of this, the hearts of all the kids are hurt. And when somebody does badly they all feel badly. Everybody feels it even if no one says anything about it. They all know if they're at the top or at the bottom or in the middle. They all know where they are in the eyes of the teacher, but it's not the truth. They are just not given the freedom to show their brilliant, extraordinary authentic selves. It is scary and it makes everybody feel separate."

Diane: *"I live with my daughter and her daughters are home-schooled too."*

R: "Home education, working with others and allowing them to learn in their own flow, is just so good.

Working with your uniqueness, not against it."

Diane: *"I was shocked to hear that Kelly might not finish given that she is moving ahead at stellar speed!"*

Kelly: *"I am a great starter that's my thing, but finishing is not my thing."*

R: "You are a starter, and that's who you are."

Kelly: *"I have somehow got to get myself to finish this. It has to happen because I would be really disappointed in myself if I don't."*

R: "That's kind of like saying 'I'm an elephant and if I can't do this just like a mouse I will be really disappointed with myself."

Kelly: *"I've just got to find a way to do this."*

R: "Or else get some help to finish it, because you have told us that you are a starter and not a finisher. That's what you have just told us. Do not expect yourself to be a finisher, so get help or support where you need it."

Kelly: "That's a very good point and I do have a good virtual assistant. So maybe I can rely on her more and more as the project progresses. Thank you."

Now to sum up.... How can you live your life, your unique expressions, to the full unless you do show who you are? To not show ourselves, to supply a censored version, is to disappoint the people who don't get to fully enjoy you. And it's living life as though you are covered with wet blankets.

I am going to include here a caveat or a warning. You must show yourself as appropriate for your own joy and safety.

My personal favourite tools are people to share and support each other. This includes: people in the flesh; people in our virtual world and through books; online videos and workshops; and networks of people offering help and mutual support.[19]

The rewards of growth and being real soon outweigh the fears. The action is self-rewarding, which is the best incentive. I am happy to leave the last words to Kelly:

"I think more people should be asking those questions about authenticity. We have a world of people that are not even aware that they're going through life not living authentically."

---

[19] See bibliography, and watch my website http://www.ReginaOrchard.com as over time more of these links will be available.

Your checklist for vibrant, abundant, good consciousness health

1. Feeling connection to all of nature - people, animals, birds, insects, plants, rocks, earth, sea, sky and stars.

2. Feeling connected to our genetic line, past, present and future.

3. Feeling connected to your own body and your inner knowing.

4. Treating everyone and everything we are connected to as we would like to be treated.

5. Knowing it is synchronicity (intended) rather than just coincidence (unintended).

6. Knowing that this is an intelligent and responsive participatory universe.

8

When you block your bad feelings, you block your good ones too!

Your checklist for vibrant, abundant, good emotional health

1. Can recognise (and accept) the sensations of many emotions in your body.
2. Can choose and use a wide range of emotions.
3. Are aware that enjoying our emotions help us to move forward, and that shame stops our forward movement.
4. Can recognise emotions arising, and then choose to release or change them to your desired emotion, rather than unwanted ones ruling you.
5. Can be open hearted when you wish to.
6. Are comfortable with change.
7. Feel empowered.
8. Feel safe in your world.
9. Can feel other's emotions (empathy) without carrying those feelings around later.
10. Can emotionally connect to others, and to the state of the larger society, without feeling like being swamped in problems.
11. Find the gifts of insights in our emotions e.g. in anger, grief and jealousy (see Nonviolent Communication).
12. Find safe ways to express your emotions so that we don't hurt ourselves or others e.g. hitting a pillow for anger and frustration, wailing out grief, art or sport to release the stress of many emotions.

Time bombs or allies?

(If you never get mad or sad unintentionally, then you can skip this section....)

We often have emotions stored away that we don't know about. They're like time bombs waiting to go off.

Unexpectedly we can become very emotional, and out of control. This can be very stressful for us, and may damage our relationships or projects. Our amygdala is very involved in this. The amygdala is part of the brain where our unconscious memories reside. They're not sorted through to make sense of them, so they are very emotionally charged and explosive when triggered. When they come to light, and become conscious, then their powerful charge dissipates

Have you ever been in a conversation and unexpected emotions appear as if from nowhere? It could be from you or from the other person. It may be joy, sadness, fear, anger, frustration, confusion or a hundred others.

Let's look at some of the many ways we can not only become more conscious of our emotions, but we can clear and release them. This way we can keep our background emotional clutter down to a "dull roar".

But first let's look at how they just appear like that.

Have you ever had a massage, and the massage therapist finds a bunch of painful spots that surprise you because you didn't know they were there? Those areas were not the issue you came to address. You weren't at all conscious of the pains there, but they are VERY painful!!

This happens all the time to us, we get used to things. We are wonderful at coping. Adjusting. We do it mentally. And emotionally. And spiritually. And in this case physically. We block out the pains, maybe put a numbness over the top to protect ourselves temporarily from the pains. But they are there, and when we're touched by another person who knows where to look, or our own touch, we can clearly become aware of what is there, for a while. And just as quickly we can return to our "status quo" - the way we are used to managing.

You may choose to allow the therapist to work on them, and allow yourself to acknowledge and release the pain as it is worked on.

Additionally you and/or the therapist may be aware of emotions that are released with the pain. These have been stored in your muscles/tissues. Some for very long periods of time. In fact with deep tissue body work, by physically working to release these, the shape and appearance of your body can change dramatically e.g. Rolfing, Structural Re-integration.

There are many ways to release stored pains, and you will be guided to work with what suits you. Over the years you may explore and use a variety of methods.

Energy healing, of which there are many types, from reiki to spiritual surgeons, usually finds no need to touch your body, though are the equivalent of physical treatments which include all kinds of massage, bodiography, osteopathy, myotherapy and healing touch.

There are mental treatments including Psychotherapy, Cognitive Behaviour Therapy (CBT), Person-Centred Counselling, Rational Emotive Therapy (RET) and methods more integrated with whole body awareness or embodiment, such as dance, 5Rhythms, Ecstatic Dance, Creative Arts Therapy, Neuro-linguistic programming (NLP) and Emotional Freedom Technique (EFT, "tapping").

Other systems for you to choose from include the Indian Ayurvedic system, the Chakra system, yoga, Traditional Chinese Medicine (TCM) including shiatsu and acupuncture, tai chi, chi gung (qi gong), gestalt and autogenous training.

Getting it right?

"If I've stopped making mistakes, does that mean I'm on the right path?" Ba Bow!! No, it means you aren't participating.

Check and see what is holding you back. Is there a fear or stress of some kind?

Really the only people who don't make any mistakes are already buried in the cemetery. Life is messy, but meaty, full of experiences and endless opportunities.

In truth you ARE getting it all right. You are either in the flow, or learning and growing.

There are light and dark times for us all

Today is a dark time, with sprinklings of light. I don't feel emotionally terrible, but my body is not well, and my usable energy is low. I feel weak, like standing up for more than two minutes is taking all my energy. I distract myself easily, but every now and then I cry like a child - as though heartbroken. And I don't know why.

Clearly I am run down. But am I emotional and drained because of that? Or did deep unnoticed emotion contribute to this stress and dis-ease?

Some looking deeper is called for. I begin to notice snippets of thoughts.

A good friend shares on Facebook that this is a powerful week (not another one!!) and time to look back honestly at our past, and see more of the story than we saw when we were going through it. To see my whole self, warts and all. Many call this the "shadow" self... parts that we tend to blame and shame. We don't want to "own" these parts.

Then the big trick, the most powerful tool for world change, is to acknowledge the whole picture and then ACCEPT IT. Realise that your whole life is "good enough". You are good enough. In fact you have always done the best that you can manage. You are always motivated by love. You are always connected to the whole, even when you didn't notice or felt alone.

You are part of the whole of life.... Consider that for a moment... Allow yourself to feel that.... an expansive, enlightening, warm and deep connection to all of life.

Stay in the stillness for a few minutes.... Breathe.... feel beyond this room, this apparent reality.... Breathe.... feel into the timelessness.... feel some of your native state.... Breathe.... let it nurture you for a little while.

As you are affected by the whole, and you cause effects, how can you believe that you are flawed and not okay?

You got here, to today. Whatever has happened along the way has taught you or others, or contributed in other gifts. Each step you've taken, each time you've laughed or loved or even just thought a kindness, you have been creating a better world. Literally.

Emotional Health

Brene Brown, in her book Raising Strong, said that we have very low emotional literacy. Of people surveyed, on average they could only identify and be aware of how 3 emotions felt in their bodies and experience.... happy, sad, angry.... Basically these three are the ones that kids learn to identify: glad, sad, mad. So in this society most of us haven't gone further. When you consider how long a list of emotions we could each compose using our intelligence, and then how many we are familiar with feeling, then there is a strong imbalance. Consider how separate we choose to keep our bodies and our minds. This is no surprise when you consider how our education system places much greater importance and value on the analytical, logical left brained activities than on feeling, sensing and more artistic, compassionate or intuitive subjects.

As I said earlier, ask any child in kindergarten if they sing, dance and paint, and their answer will almost always be "yes". Ask any adult and the answer will be very different. Our society has developed many blocks and challenges to our natural expressions. (See Playfulness in your life).

Whole Brain working together

Let's have a brief look at Left Brain and Right brain learning.

To help remember: Left brain, L is for logic. Right brain, R (ar) is for art.

So in more detail: LOGIC side is used for sequential or step-by-step learning, Language, Arithmetic, Analysis, Separation, Audio/Sequential processing etc. ART brain is used for WHOLE concept thinking. Connection, Art expression, Mathematics, Imagination, Intuition, Visual/Spatial processing.

The right brain is a full conceptual brain and has been used for thousands of years by traditional, holistic and integrated cultures.

Being able to see the big picture and new solutions is all through the benefit of the right brain. Using our right brain has been downgraded in our Western education, and in what we value.

So I encourage you to develop the creative side of your brain which taps into your inner knowing. I will give some guidance in to this, but it's not rocket science, it is surprisingly simple.

Our education system and workplaces value LOGIC over INTUITION. Einstein stated that imagination is the king and logic the servant, whereas we clearly have it the other way around.

This paradigm has taken us to an untenable world. Most feeling people are dissatisfied or disillusioned with our current direction, and I can agree for the Western world, because that is what I'm familiar with. There are huge problems in most sections of society, from environment to education, health, equity, governance and many more.

If we want to improve our world and design better solutions, as Einstein has said so well, "You cannot solve a problem from the same consciousness that created it. You must learn to see the world anew."

I am putting the case throughout this book that it is prudent and worthwhile for us to change our consciousness and so see our world anew.

And I hope I'm sharing some tools for living with a connected and whole approach to life.

The goal is not to use just one side of your brain, but to use them both together, with intuition as the leader. Doing exercises in this book will have you spending more time connecting with your whole brain and body.

As with hands, if you are dominant with one, and wish to change that, you will have to go through the discomfort and awkwardness of using the other hand, and building your skills and stamina with it. And as your abilities grow you will feel more trust and become more willing to use it for a new purpose too. Be kind with your whole brain. It's powerful and will become great fun, but be gentle with yourself, because in this human existence things do take time.

"Being" in your heart or your head

The difference between "being" in our heart or our head is huge. It affects so many areas of our lives. When things are going grimly, we're pushing ourselves or are too stressed, or there is lots of self-sabotaging self-talk, then you can bet that the head is in control.

The wily mind is like a bucket of slippery worms - all intertwined and ever moving.

A better result is to come to tasks or relationships, leading with your body. We all perceive uniquely, so it may feel as though it's your heart, or your solar plexus, or another part, or perhaps your body as a whole.

The head certainly plays a role. It assists with planning, organising, analysing, and taking action.

But the wonderful warm feeling that is in your heart or body, sometimes even feeling so deep and precious that tears well in your eyes, is a clear indication that you are connected to the flow of the universe. You are tuned in, you are on the right channel, and the ideas and words coming to you are the most constructive and beneficial ones that you can be using.

When you're having a "tough day" or a "shitty moment", how do you get yourself back from your head (thinking) to again lead through your heart (feeling)?

Here are a few methods. (Also see "Small Steps are Powerful Checklist").

gratitude lists	laughter
time in nature	playfulness
favourite exercise	hugs
gardening	love-making
playing with kids, dogs, or other animals	a project you're passionate about
go on a date	art & creativity
meditation	energy work
breathing	mindfulness
share a meal	walk
be with friends	dance or move

If you're like me then setting up a whole list takes too long. So you can copy and paste this list into a separate file, or print off this part, and mark on it what appeals to you. Create your own first aid list, and put it where you can see it. You are starting to create a new habit.

Now please stop reading and go brainstorm to create a list of your methods. Ones that you already prefer to use, and new ones that you'd like to add in.

Great, you have your first aid list, which you can put in several places to help you. When you're going off into your familiar old pattern, quickly grab that list and do any one of those things RIGHT AWAY. It will help keep you on track. Even if it's just for a minute or two, it will get you back into your body, into NOW.

Prompts are very helpful. Your memory can be kick started by mentally tying some of your first aid items to objects or other activity in your surrounds.

Example 1. Do you have a heart symbol visible? Perhaps it's in a decoration, jewellery, artwork or a card? It's a useful reminder to do some heart breathing.... Sit for a few minutes looking at this symbol and feel your heart open and warm. Focus only on your heart, the warmth, and imagine you are breathing air in and out through your heart. When you finish, notice any differences you feel, including sensations.

Now for one more minute focus on your intent to use this heart reminder when you are in your head and consequently feeling discomfort. Focus on how you feel now that you are back "in your heart". Intend to notice the heart whenever it will be of help.

Example 2. Water is a symbol both for emotion and cleansing, a great combination. So each time you have a shower, create a ritual to check how your chakras feel, and to imagine all the yucky, grey energy pouring out of them and being washed down the drain.

Children's values

Primary (elementary) school children tend to understand what is real and what is important.... Most of them intuitively feel connections to our world. It is worth observing their wisdom while they are still very connected within themselves and to their world.

From observation I feel most young children stand for these values: kindness, treating other people like you would like to be treated; it's ok to make a genuine mistake, but important to do your best to tidy up or make amends for problems you cause others; sharing, and how it's not right to treat others badly so that you can get ahead. They also understand about connections such as the water cycle, and how drops of water travel all over the planet, and how adding poisonous substances can eventually affect any of the living things on the planet. They understand that life is much better when you have friends to share it with, and if you treat them kindly they are more likely to want to stay friends.

In most classrooms there are also a broad mix of children, and teachers often spend quite an amount of time helping these diverse children understand that we are all a part of community, and though we aren't all close it's still important to find ways to live together, and to allow everyone to thrive. Sometimes this is the hardest thing to do in a group, because we can feel so irritated by simple things that people do.

But in school most children will have these natural tendencies.

Children live their wisdom. We can observe them, to help us remember our connections and what we value.

Your Inner Child

What would YOUR inner child say?

As adults we often censor ourselves. We often don't allow ourselves to be our best / be real / care / put ourselves on the line. Children don't know all the rules, and respond honestly and openly. They have a direct line to their hearts, until eventually they learn that it's not safe. Then they begin to barricade them.

But that child self never leaves us. It might be pushed down and ignored, but at every age you always have a child in you. How is your inner child? Do you notice it?

Yes, you have one. He or she may be various ages in various aspects of your life. You can recognise this child part by the high emotional level, and the often out-of-control nature. It's where your pains are. Your playfulness and fun are too. It depends how healthy this child is.

I can tell you a bit about my child. She wants to be able to see the fairies at the bottom of the garden. She loves to have family and friends to be with, play and hug. Loves laughing, eating cake and dancing (well she calls it dancing - not everyone would). Gets worried about paperwork and taxes, and likes to stay up late when I'm not noticing. She happily prioritises fun over essentials, and often with her enthusiasm she will get the essentials done too. She loves holidays in warm climates and will exercise, and enjoy it, if it's with friends or if it's fun.

I've helped her grow up a bit over the years. She used to get very stressed and cranky when she felt things were getting out of control.... when she felt she/I wasn't being a good enough mum, or when feeling ashamed. One more thing that I get in to trouble for is when she uses superlatives.... fantastic, the most wonderful.... it's not accurate enough for most adults.

Tell me, when does your inner child show up?

Look into yourself, or into a mirror, and when is your heart open as wide as your eyes, and love is all there is? Feeling kindness and warm like cuddles?

That's your precious inner child clearly present.

And I'm sure you sometimes find yourself upset, or fearful, maybe unreasonable, digging your heels in, blaming someone else, wanting to be right instead of loving. That's when your inner child is taking over, and instead of informing you, it's in control, and is not being a friend to you.

This is time to be the compassionate parent to your child,[20] to be gentle and understanding while keeping that child from doing damage. As Brene Brown explains, to allow that pained part to be there, as you want all parts of you present, but don't let her take the driver's wheel.

Would you like to be more in touch with this precious child? Here is an exercise to do:

1. Have a conversation with your young self, or your inner child, however you wish to think of him. You write with your dominant hand (right if you're right handed), and he writes with your other hand.
2. Talk to her. Ask her questions. Have a conversation. Remember to breathe as you write.
3. Or go somewhere he might like and talk to him, either silently or out loud.
4. Remember that children need to be made to feel safe, protected and appreciated. They respond best through the heart (and to playfulness and imagination).

Bonus question: For a minute, considering the state of our world, let's remember the primary schoolers, as they have clear wisdom.... If you ask them which is real and which is most important.... society or environment or economy.... do you know what they would say? What would your inner child say?

---

[20] See Transactional Analysis to help understand and deal with our inner adult, parent and child.

Ask a friend for what you need - for your inner child

I wrote this book to a deadline, and one evening, when my partner had been away working for weeks, and I'd been working through this unknown writing and re-writing process, I was feeling frustrated, unsure, a bit overwhelmed and very much wanting physical closeness and some reassuring words for my inner little girl).

So I texted my darling man, asking for a book "pep talk" phone call, which he willingly gave. He listened, we tossed around a few ideas, and it helped. At the end of the call I said "Please, would you imagine I'm 3 years old, and I really need some encouragement and kind words."

In a gentle, tender voice he said

"I would run my hand down your back,

Pull you close to me, and whisper in your ear

You Can Do It....

Nothing is impossible."

His simple reply brought me to tears. I felt understood, acknowledged and loved.

And after that I felt so much better. I thought of it a few times that evening and in some days after, and I felt a similar response each time. And somehow my creativity and productivity rose too.

I encourage you to #askforwhatyouneed

Exercise - Your Young Adult Self

Here's an exercise to explore your life as you were entering adulthood. You'll travel to the time when you left school and as you were going out into the adult world. (It is better if I don't give you goals other than getting to know yourself more, and that is enough reason)

This is a process to explore your embodied wisdom.... and yes you do have much inner wisdom. Often we're afraid to look because some of our depths are unknown territory to us. But rest assured that you WILL have gems of wisdom. You are a person worth getting to know better, even if you believe otherwise. Conversations with your deeper self will give you rich rewards.

I am in awe at the self-loving insights that people gain from arts therapy exercises. So I felt I had to include one in this book. If this doesn't work for you through the book, then join me instead in person or online for a more immediate experience.

I'm going to explain about this very briefly and then experiencing it is the main thing.

You'll need your arts journal or some paper, a pen and some wax crayons, pastels or watercolours - a simple, easy way to put colour on your paper in an uninhibited way.

The goal is to give you an embodied experience, so there will be various times during this process where your brain may want to say "No I don't want to do this! This is too hard! Leave me alone! I have better things to do," and all of that.

So just know this is coming and give yourself a hug and keep going, because those will pass, and at the end of the process you will have a different perspective.

Please, know that you are in control. Only do what feels right for you.

Journal, pen, colours and cuppa in hand? I will give you just enough explanation to do this; that is part of the process, but it will be enough. Read one step and do it before reading the next.

Ok, here goes:

1. How were you by the time you left school? From the following words select any that applied to you, and add any others that occur to you.

   satisfied, joyous, relieved, fed up, disillusioned, on top of the world, supercilious, motivated, inspired, prepared, cowed, unprepared, lying through your teeth, feeling like an impostor, humble, tired.

2. Record them in your journal.

3. For each word you choose, allow yourself to write down associations in each of these 4 categories: thoughts, images, feelings and sensations.

4. From all these words, select key words - the ones that you feel more strongly than others, whether you feel attracted or repelled.

5. Staying with your feelings, express them as colours and movement on a page. Allow your body to move freely to express whatever feelings have arisen from this process, and intuitively make coloured marks on the paper until your feelings are released. Focus on the body, not on the mind, and allow it to bend and move as IT wants.

6. Add any more words that arose while creating your image.

7. Use all your key words, and any other words that you wish to add in, and let yourself create a poetic statement of any length. (Keep going, don't think much, just keep going, keep breathing and let it happen.)

8. You managed to go into the unknown and do this strange new process? Please congratulate and appreciate yourself and your adventurous spirit.... Every new step is expanding your world of possibilities.

9. Sit with your creations. Leave them where you can see them, and try to spend time looking at them throughout this week. Don't think too much, just absorb and appreciate.

Write a letter to yourself from a different perspective

This is a very easy exercise, because the main idea is to breathe, feel and write.... not much thinking. A pen that you enjoy using, paper or your journal, breathe and begin.

Here are a few suggestions to inspire you. Select one of them and write this letter:

- From your loving, open heart, or
- As a loving parent to a young child, or
- From your guardian angel, or
- From the divine web, the Great Spirit or your loving God, or
- From your 8 year old child self.

When you've finished, read it back to yourself, and really be present.

E.g.: Darling Regina,

So how are you doing today? Are you in touch with your feelings? It's been a couple of rough days, and I'm so pleased to see that you are getting back into your book. Doing these 5 pieces of writing is great. Sure they need some editing and tidying up, but it feels like good stuff doesn't it?

Yes, a bit slower than you'd hoped, but an extra minute here or there is no problem.

I just want to tell you that the core of you is very pleased with your work, and that moving forward with love, from your core, is the best and most valuable work that you can ever do.

So thanks, and keep up the good work. Much love from your core, and the universe.

9

Feeling yucky is just a habit

Yucky (comparative yuckier, superlative yuckiest) 1. (Colloquial, often childish) Of something highly offensive; causing aversion or disgust.

Yucky feelings are just habits!! Let's see how it works.

"I can't rely on myself. I let myself down!" Mary wails for what must feel like the zillionth time in her life. The thing is that Mary CAN count on herself. She is very reliable. Her feelings of low worth and unreliability and her worries and fears about herself are totally predictable. They are a bunch of habits.

Emotions give us feedback to our previous behaviour.

Emotions are the markers, both consciously and unconsciously, of what we think, how we feel and how we have acted.

The emotions are results of our habits....

If these habits are in alignment with our true life direction, the results will be feelings that are comfortable for us, and are likely ones we call pleasant.

If these habits are not in alignment with our direction, it is as though a big NO response comes rushing at us! It's a yucky feeling to get our attention, to prompt us to notice and hopefully make a change in our patterns. (In a similar way that nightmares are dreams trying to get our attention to look deeper or change something that we are doing in our waking life).

We usually have long tern patterns of YUCKY FEELINGS - the ones we don't enjoy.

They stop us and block us from our goals and intentions.

Because of them we may feel "I'm not good enough" or "I can never get it right" because "I feel stuck, I feel yucky, and I can't achieve what I hope to achieve." This cycle tends to be self-perpetuating.

So in some areas of our life we may feel STUCK IN THE YUCK!!

BUT SCIENCE has now confirmed wonderful new insights which we can apply to change all that!!

Let's look at emotions differently.

How about we look at emotions as a warning signal or sign? A warning that something we did has created that yucky emotion.

The emotion showing up means that something earlier in your chain of actions and thoughts created that. And you have a choice to change those things, and instead create a feeling that you really like.

With the yucky emotion, chemicals pump into our bodies and strongly reinforce it.

In fact Dr Joe Dispenza in his book *Breaking the Habit of Being Yourself* has outlined the research which shows how the body is doing the equivalent of hard wiring those unwanted emotions into her personality, using chemicals. Mary's behaviour is the equivalent of a cocaine addiction. So it's totally reliable. And unlike a cocaine addiction she has no idea of the cause, and simply blames herself for failing YET AGAIN.

Now for the good news

It's important to know that neuroplasticity is one of the discoveries that states that we can add new pathways in our brains, in effect new habits, easily at any age!!

Let's emphasise AT ANY AGE!! So forget the old adage "You can't teach an old dog new tricks". Because science has proven that old and young can learn easily IF WE KNOW HOW.

We create these pathways by what we think, what we imagine, and through our emotional state and our senses.

Also, perhaps even more exciting: we can also RELEASE PATHWAYS we no longer want!!

It takes awareness, it takes some time and focus. But considering the effort required to change the habits, the PAY-OFF IS HUGE!! It can feel like winning the lottery. Your life becomes easier, more... (Fill in the blanks of what you'd highly value)... easy, abundant, joyful, harmonious, successful and loving.

Now it's been observed that most people have a lot more yucky rather than preferred emotions. Consider if this may be true for you too. This can inspire many of us to take the steps to make the changes.

I'm opening this up so we can all be aware that this is a simple, definable, and doable process.

Now I'd like to take you through a simple exercise. It's a memory exercise, but I want you to remember it in your emotions, in your senses in your body as well as in your thoughts and in your mind.

So please just sit there for a minute... Notice your body.... take a couple of breaths... perhaps close your eyes.

Now I'll give a couple of examples, and perhaps a moment of your life will come to mind. A time when you were feeling yucky, whatever that was for you. And then something happened and almost instantly the yuckiness was replaced by something quite different, a more life-enhancing feeling.

Perhaps it was a dear friend or family member calling, or you found out you'd won something, or you discovered some good news.

Now when you're ready, briefly allow yourself to feel that yucky moment....

Now notice and feel that moment when how you feel changes so dramatically....

Feel the contrast....

And now please return to the present....

Take some time to record this for yourself. It is important to be able to remember that you already are able to change so quickly.

I wanted you to feel one of the many moments in life where you have been able to change, almost at the flick of a switch. We can harness this natural flexibility. And the first step is to begin to realise that we have the ability. Now it's a matter of learning to do it consciously whenever you choose.

I just want to remind you how remarkable you are, and if you can do it in some things you will definitely be able to do it in other things. All these abilities are totally transferable.

Awareness is half the solution, and then there are many methods to make the change. (See Brain Watching, Brain Washing and Living in your desired life). (Dr Joe Dispenza, *Breaking the Habit of Being Yourself* and *You Are the Placebo.* Esther and Jerry Hicks)

Appreciation or gratitude experiment

1. For 1 hour choose to be grumpy and blame other people as much as possible.

2. See how that feels.

3. Write down some key words of that experience (how you felt, your sensations and/or thoughts).

4. For the next hour choose to be happy and understand that you have the power of choice - it's not other people's fault.

5. Observe how that feels.

6. Write down some key words of that experience.

7. Compare and consider these two different experiences and create your conclusions.

This is well done when out in the world interacting with others.

Less intrusive and for quicker results you could do this for a short amount of time such as 5 or 10 minutes.

Special note: If you have wondered what the point of this exercise is, that suggests to me that you will benefit by doing it.

I can almost feel your response: "Oh no! That's certainly not what I wanted to hear!"

Here's the deal: if you don't understand more after doing this, you can write me grumpy messages... I'd actually be interested to hear.

Checklist: from yucky to great

1. Plan out ahead how to deal with these interlopers while you are at your best.

2. Rituals are very helpful to use to create new habits.

3. It is important to say "This yucky feeling is only a habit, it's no more real than anything I have created. I am learning to create what I want." This brings you back to being a creator. It returns your power to you, so right now you remember that you CAN change your pattern.

4. (See Brain Watching, Brain Washing, Adding In and Fading Away, and Living In Your Desired Life)

10

HELP, the world's not going the way I expected

Common questions from activists

From activists, these are common questions.... "Why do we work so hard and still are not getting the results that we're working for?" "Why do I seem to be hitting the same obstructions in different projects?"

The energy shapes the results. The energy is affected by our physical energy, our emotions, thoughts and beliefs, including all the unconscious ones. Energy flows where our attention goes, and that's also energy from others. It's one of the laws of creation in this intelligent responsive universe in which we live. It is a Universal Law and we can test this out in many ways. It does require us to be honest and to look deeply to see where our attention is really going. For example if we are putting attention on having something new in our life, but at the same time really noticing that "it's not here yet", then we are working against ourselves.

Esther Hicks channels some wonderful information which assists people to understand more deeply the laws of manifestation and how to bypass our minds which are like a can full of worms. Lookup 'Abraham Hicks'.

The Buddhist and Zen masters, Native Americans and the Gnostics also state the importance of being in a state of already having what is desired. And according to scholars including Gregg

Braden and Catherine Ponder,[21] the Christian Bible also contained such references in the early versions before editing and further translations.

Any feeling of lack will attract the state of lacking. And that is why we often get what we don't want but we can't stop paying attention to.

Have you ever noticed the times when you were focused on other things and what you really wanted just happened, just turned up so easily? This is a situation that you wish to duplicate. Get busy creating things so that you forget what you don't yet have. Also daydream and vision what you are creating as though you are already living in it.

In a group it is great to enlist your day dreamers to this project, and get everyone else totally distracted or focussed on achievable outcomes. Consider this as a strategy.

Now don't worry, because after you become aware of these issues, which by definition increases your consciousness, we will explore solutions that we can all apply. You will be able to find methods that suit you. You will be able to consciously choose what you desire, and change your mind and your targets more freely.

More exciting experiments

I'm excited that at last scientists are proving wisdoms and faiths that they have previously disparaged.

Like most of us reading this, I've spent my life looking for the answers I needed. I'd found some great ones, but I hadn't been able to put them together to be able to fully answer my needs. The pieces include: a lot of traditional wisdom from diverse cultures and belief systems about how the universe works; how to be productive

---

[21] See also writings by Myrtle and Charles Filmore from the Unity church, and the Metaphysical Bible Dictionary.

and enjoy life as much as possible; how to be healthy in body, in mind and in spirit; how to share these understandings with others; and how we can create a sustainable, loving, fun and peaceful world.

I have found that when we challenge the culture we are raised in, then confusion, doubt, blame and shame all arise for us to deal with. But that's okay. We have good tools to deal with this, and now there are millions of others going through it with you. The internet is full of verified experiments, explanations, tools and support.

Yes, it is wonderfully exciting! But I'll calm down enough to continue.

This next experiment[22] was done in 1993.... The researchers took[23] a glass container and created a vacuum in it, and then, as they knew would happen, they were able to measure the photons still in the container. Photons are tiny, tiny sub-atomic light particles, which are the matter which comprises the universe. All the photons in the tube were in random positions, not in any particular shape.

Then they took some DNA from a person, and added this DNA into the container. To the amazement of the scientists, the photons then formed into patterns. So the DNA from the human body affected the stuff that the universe is made from!!

They then expected that after removing the DNA, the photons would return to the random form. But no, they took the DNA away and the photons stayed in the pattern. Initially this result was inexplicable.... Stay with me, we will get to how this all fits together.

---

[22] P.P. Gariaev, K.V. Grigor'ev, A.A. Vasil'ev, V.P. Poponin, and V.A. Shcheglov, "Investigation of the Fluctuation Dynamics of DNA Solutions by Laser Correlation Spectoscopy", Bulletin of the Lebedev Physics Institute, no 11-12 (1992): pp. 23-30. (for google search: DNA Phantom Effect)

[23] Vladmir Poponin, "The DNA Phantom Effect: Direct Measurement of a New Field in the Vacuum Substructure," performed by the Russian study again in 1995 under the auspices of the Institute of HeartMath, Research Division, Boulder Creek, CA

This relates to and corroborates Dr Emoto's water research where people's words and intentions profoundly affect water. The water droplets are then frozen and the patterns of the water crystals change. Those that have been prayed over or labelled or spoken to with love, peace, kindness or similar qualities form regular and intricate patterns of great beauty. In contrast the water that is abused, labelled as hate, etc., or had certain harsh sounds played to it, forms irregular crystals which are unattractive with no clear patterns.

This is another validation of the power of intention or prayer. And it also gives a lot more detail about how to do it. According to scientists it's not the prayer as such, or rather what the prayer is.... it's the emotion of compassion or love.... and the intention.

An often repeated experiment, of a particle which can change between a particle and a wave, depending on the observer's expectation, is yet another example of this being a responsive or intelligent universe.

Remember our bodies and our earth are about 67% water. Yes the water in our bodies also responds to intention, and our cells do too.

Basically the results of these experiments mean unimagined possibilities. Our goal in the book is to explore how some of these apply to life here on earth, and pointing you to a variety of concepts and actually give you tools to use each day of your life, if you are drawn to any of them.

In a nutshell:

So many scientists are now saying that there is a web or a matrix of something that we are all part of. We are all connected into this web. We are instantly connected. What anybody does in this world has huge effects. And our thoughts, our intentions and our emotions very much affect our minds, our bodies, and our spirits. And not only our own, but those of many others.

Of course there are people of all kinds with great personal power who have cultivated their minds, consciousness, and a loving vibration or presence, e.g. Mahatma Gandhi, the Dalia Lama and Nelson Mandela. This people create huge effects. I'm sure you are aware of inspiring people, look at them. We can all choose to be inspiring in our own unique ways, and we can learn to increase our personal vibration, and hence our creative effects on our world.

These discoveries legitimise and validate many methods and much wisdom from indigenous and traditional people, healers, psychics, and sensitives, etc. They are now not only seen as possible by the "logical community", but many things can now be measured and verified. This includes healing in person and at great distances, telepathy and other intuitive understandings.

Remember though that science is not the arbiter, as science is the study of life. That means that life and its occurrences and discoveries come first, and then science observes it, and works to understand what already is. It has taken thousands of years for science to be able to see this much.

Do not wait for science to verify and approve of what you do. However be consoled and supported that the logical view can now see more of what is really happening.

Do you wish to be on the cutting edge of creating your part of your best possible world? Then try things. Observe yourself. Observe the effects and connections. Explore.

The story so far: a summary

> *"Though the problems of the world are increasingly complex,*
> *the solutions remain*
> *embarrassingly simple."*
>
> - Bill Mollison

For many of us there may seem to be a missing puzzle piece when all of this is put together.

If it's all crystal clear for you so far please skip this last section in this chapter, or if you'd like a bit more grounding please read on.

Recently, especially in the last 30 years, science has reached the point of having equipment sensitive enough to be able to find out what's happening at a really basic and sub-atomic level in our universe. This has blown them all away, as the discoveries diverge so hugely from our accepted views. The scientists have had such huge shifts: from genetics to epigenetics; from Newtonian to Quantum physics; from "you can't teach an old dog new tricks" to brain plasticity and psycho-neuroimmunology; from "psychosomatic illness" (it's just in your mind!) to "there is connection between every part of you and every part of this world" and we affect each other.

As I said in chapter one, for whatever reasons, not many of these life changing discoveries are included in, or are affecting, our educational curriculums at any level. They're not being promoted. They're not being mentioned much in the media. (Another reason to avoid the mainstream media).

The ramifications are extraordinary, and I've based the book around these discoveries.

I am giving you brief examples of experiments that can change the ways we choose to live our lives. All the research, the scientific proof, is readily available.

We have techniques, awareness, consciousness, talents and tools. We may even have a fair amount of faith that if we put these in place then they will work. Usually, for one reason or another, we are somewhat dissatisfied with our life results. If we rated our life satisfaction or frustration, there would be a gap between our expectation and our perception of reality.

We also have to deal with our past levels of effectiveness, where we may be judging ourselves harshly - we just didn't perform as well as we'd hoped, perhaps we dropped our bundle, or didn't achieve as much as expected, or completed but felt weighed down emotionally, etc. Often we are doing our best to work on living by faith rather than certainty.... Though you could argue that no matter what you do, life is pure guesswork!!

Well I am being bold enough to suggest that we can step from faith (or lack of faith) or hopelessness, to more joy, satisfaction and certainty! (Well at least as much as we are ready to allow ourselves.... After all life is a life-long process!)

Here is a puzzle piece for you for free:

A previous section explained the scientific proof that this is an intelligent responsive universe!!! That we, all life on this planet, and perhaps in the whole universe, is immediately interconnected. We just need to learn how to operate the wonderful intricacies of this!!

As well as this small point, (ha-ha - a bit of joking here to help ease the shock factor that you're likely still dealing with) here's a second one:

It's not so much that we're people who might choose to be spiritual. Really, evidence is piling up that we are vibrationary beings or consciousness living with or within our bodies. The point is we all are drivers, and we're not just our bodies. We are much more capable, creative and powerful than most of us would care to guess. I'm writing this to help you see and understand some of the control panel in your cockpit, and where your virtual jet aeroplane, or even spaceship, will take you.

Metaphorically speaking, many of us think that we don't even have a virtual bicycle.

Then we need to realise there are many different systems which explain and make meaning of this:

Here are just a few: angels, God (with many names), gods, deities, fairies, plant devas, prayer, karma, dharma, consciousness, energy, power of love, Gaia, psychics, string theory, healing, meditation, the quantum field.

Remember how scientists are proposing that the universe is made of some kind of base material, like a web that weaves through everything and everyone in the universe, even in the so called vacuums of space? Yes, this seems extraordinary if we haven't been taught to consider it.

When we start to think of all the synchronicities (apparent coincidences) that occur in our lives, we can start to realise that we've accepted them without even the acceptance of a web. And that the web fills in a missing piece of our understanding. All those times when the phone rang and you knew who it was before you answered the call (before caller ID was available). And when you did something on a whim and you ran into just the right people, or made a great contact, or saw something that helped you or a friend. Now we can see the medium for many of these things. And we know that our consciousness, or our intention, creates results.

In fact Henry Ford's "Whether you think you can, or you think you can't, you're right" takes on even clearer truth. Our thinking, vibrations and expectations affect the results, affect ourselves and others, and our environment.

11

Working with others ... from hiding to thriving!

All our living is a chance to draw in the wonderful energies around us... especially from:
- nature
- joyful events
- art
- our achievements and creations
- young children

Babies and their births are such wonderful events. Many people love to enjoy beautiful energy from being with young babies, children, animals or nature. It makes perfect sense because plants, animals and young children are all safe to be with. They are not going to be hateful, hurtful, violent or repressive.

Energetically they will share yours and theirs, easily, perhaps playfully. And all parties will feel better after the exchange.

Two People Loving

It's a great truth "that what we put our attention on grows", whether it be tending a garden or a business, family, lover and partner, friends, or special projects. When we attend to it, which in French means "to listen", there is a blossoming.

Sexuality is a sacred sharing. What does sacred mean? Something that is precious and to be valued, respected and protected. It's also something bigger than us, or is us at our best... our divine part. Two people can create energy that beneficially affects the world.

So for me sexuality is about bringing two people together for that divine merging. It is an opportunity for two whole people (hahaha.... or fairly whole, the best that we can manage at the time.... perfection is not our goal). Our goals are both giving and receiving of these: joy, love, being in the moment, openness, kindness, support, pleasure and bliss.

Tantra is a study of merging energies, of coming together and creating energy to share out in the world. Taoist Sexuality practices are of a similar type. (See the many books of Taoist practitioner Mantak Chia.)

What does sexual intimacy have in common with time spent with nature and children? They all include deep levels of: trust, safety, connection, and natural flow. Each being is open to the other. This is a powerful creation for two adults. Especially if we've had cause to mistrust in our lives. And this way of couples being, with unconditional love and connection, is a great step and a great healer.

From I to We

We are always creating our world. We create where our attention goes. We may focus on our clear intention, or we may focus on what we are against. We create what we focus on. The world, at a sub-atomic level, responds to our intentions.

Oh bother!! ....Yes sometimes it goes very differently to how we imagine.

But the point is, we've done our best till now, and it's also had benefits for others that we've never guessed. So no point beating ourselves up!!

Take a breath, shake it out. Laugh at the silliness and the divine/sublime beauty of this unimaginably wonderfully designed system.

Be in awe. Then realise we each are totally connected, and are pure creative potential.

It's okay if you don't believe me, that's not the point.

It's taken me decades to arrive at this viewpoint. And I'm delighted and shocked that many younger people seem to be arriving here before me! But take credit in all the steps and awareness that you have achieved yourself.

And bask in the support from your friends and other loved ones.

Support is a powerful force. Without this force I would still not be writing this book. I probably would be yearning to, but not feeling "enough" or "legitimate enough" or "clever enough" or "wise enough"... there are so many "not enoughs" that we are used to being. Habits get set in place, and we tend to accept them without questioning.

A network of live, passionate people with big hearts.... that is perhaps the world's most powerful weapon.

I wrote that and then recalled that even the military around the world have indeed done lots of research on people's potentials, and have realised that the powers of humans are so important to explore. So far we are still only tapping a small percentage of our potential. Einstein estimated that we are only using about 5% of our brains. Perhaps some people push that to 10%.

For me a major benefit of the work is more feelings of peace, and of a heart-opening joy of being alive, and how this spreads to others with positive effects.

When each of us show our hurt "inner child" or "pain body"

There are times when you behave like a childish kid, not like a mature adult. There are various names for this, including your inner child, your hurt ego, your shadow, your pain body, your unconscious self. Our buttons have been pushed and this part rears up. We may be self-righteous, confused or embarrassed by our own behaviour.

So what do we do?

Here's my recommendation for when we're acting childishly:

1. Be kind to yourself.
2. Be willing to accept these parts of your unique self.
3. Forgive yourself and remember it's part of being human.
4. Remember the fullness and brightness of who-you-REALLY-are (the soul, part-of-the-web-of-life, connected-to-all, a creator).
5. Move forward.

Now remember the last time someone close to you had one of their buttons pushed and they acted like a kid (or call it their inner child, hurt ego, pain body, unconscious self, etc.).

They are just like us!

So what did you do?
- Understand?
- Get cross?
- Not take it personally?
- Retaliate?
- Laugh?
- Take a deep breath and maybe a short walk?

If this adult can have their scared child treated well by you, this becomes a healing moment for you both. This is a precious opportunity with your most tender parts. You are building intimacy of caring and trust.

We all have this side, and we all yearn to be treated with kindness and respect, even if we're out of control.

Dan Siegel in his book *No-Drama Discipline*[46] offers many tools that work for adults as well as for the children in a family.

When someone we love is acting childishly:

1. Remember who they REALLY are, and appreciate them.

2. Understand.

3. Forgive their actions.

4. Don't take it personally - they are on automatic stress pilot - they are out of control.

5. Treat them with kindness. In kindness they are more likely to receive the message of who they are.

6. Accept the whole person. Do not expect them to be infallible, as no one is.

Acceptance and kindness goes hand-in-hand with self-care. It's also healthy to calmly and clearly state your boundaries.

For example, you may be committed to turn up and work through such situations in a relationship. But you can make it clear that both of you need to be keen to do this.

One person can't keep a relationship healthy.

---

[46] No-Drama Discipline: The Whole-Brain Way to Calm the Chaos and Nurture Your Child's Developing Mind: Daniel J. Siegel, Tina Payne Bryson

Your Greatest Ally

There is one other person who we are fortunate to have as a potential close ally. For many of us this ally does not seem close. The intimacy has been long neglected, and we may not even know what they like, want or need any more. This is because we are out of the habit of spending time with them.

This major relationship is the one that you have with yourself.

Is this a good time now for you to stop reading and spend a while "attending" to yourself? Basically you will be having a catch up over a cuppa with a dear friend. YOU.

So while your kettle is boiling for a favourite cuppa, find a pen and paper, or start up your computer. Treat yourself with a yummy snack as you would for your other dear friends. Find a place to sit that feels good to you.

Have you taken a few minutes to do these small loving kindnesses for yourself? Remember Kindness Is Small Steps. Small moments build into life affirming habits of self-nurturing and insights.... Great, you're comfortably settled and ready to chat with your greatest ally.

You're going to write quick answers to some questions. But before you begin writing, become aware of your breathing, and then take 3 deep breaths to help bring your attention here more fully. If you are able, breathe in through your nose for a count of 4, then out through your mouth for a count of 6.... Continue that for two more breaths....

When you're ready, allow answers to each question to appear with little thought. Ask yourself a question and wait expectantly for a response. Just write down whatever comes to you. Try not to judge or edit, so that you will tap into your inner knowing.

(Sometimes ANYTHING needs to come out first for a while to get to it.... so no editing.... just ALLOW.... it may not be polite or logical... allow yourself to rant, rage, complain.... whatever it is, welcome it).

1. When is the last time you sat for half an hour and asked yourself questions?
2. Do you feel very in touch with yourself today? This week? Not for a while? Not for a long time?
3. What is your current biggest goal?
4. What is your biggest worry at the moment, and in which part of your body do you feel it?
5. Describe its colour and shape and any other details.
6. When else is this feeling present?
7. What other feelings do you notice right now?
8. What is one step you could take today to bring you closer to your major goal?
9. Is there anything else you'd like yourself to hear right now?
10. What pampering would you like this week?

Right, you have just taken some time to check in with your most precious ally and friend. She/he is a great confidante and holds much embodied wisdom to share with you. If you take time regularly to dip into this resource you will feel strengthened and surprised.[24]

---

[24] Resources: The Artists Way Julia Cameron (and morning pages and artists dates), Arts journaling, Expressive arts, journaling, intuitive writing, intuitive dance. In fact almost anything that is intuitive. Travelling alone.

I am the one

"When we choose to accept responsibility for who we are, what we say, do, feel and be, we are ready to take control of our lives and our future." - Pamela Harrison[25]

There is no blame for myself or others. But everything is something that I can do something about. When something bothers or upsets me, it is another chance for me to remember who-I-really-am and make a change. I'm the one to do the work.

Perhaps:

1. I need to set my boundaries when other people are behaving badly or trampling on me.

2. Each time I feel "less-than", there is something I can do, because in truth I'm not "less-than".

3. I may need to forgive myself or others.

4. There is something for me to learn.

5. There is something for me to let go of.

6. There is a new understanding for me to integrate.

This may seem pretty overwhelming and like a lot of work. But this is truly empowering.

I am not powerless. I may be in the habit of seeing myself as less, but I am much, much much more.

I am the one. I am connected to all. I have effects on all. My intentions and actions affect others. I am powerful.

---

[25] Pamela Harrison http://www.thenextstepcoaching.com.au/videos.php

When I feel that another is a problem, and I realise that they are being a mirror for me, then I can change myself, and my vibration changes too. Then the "problem" person will either change or go elsewhere. If my vibration no longer matches theirs, they will either let go of their problem, or leave me alone.

I need to remember I don't need to do all this alone. I have friends. I can ask for help. I find people in my tribe. The universal web brings wonderful assistance when I allow it.

In fact there is a universal truth that we must ask for what we want. So spend time writing and saying out loud what it is that you do want in your life. Ask for:

- Help
- Support
- More Ease and Grace
- Abundance
- Wisdom
- Solutions

And after asking, also remember to relax enough to see the help when it is offered, and to welcome it and invite it in.

Importantly spend time imagining that what you are asking for is already in your life. Imagine how you feel, including all your senses. You can also find images to represent your goals: create vision boards and treasure maps (there will definitely be "how to" guides for these on the internet, including on YouTube).

When we ask for our dreams to easily come true, we are calling on a spiritual truth, allowing who-we-really-are to be expressed.

Note: When I say "easily come true", as you can see by now there are a lot of factors. I have shown all kinds of resistance on my wiggly life path. I have felt pain and I've suffered too, but I am learning! As are you! Many of you will learn faster than I do. It's a journey, so as we become more loving with ourselves it will be easier, and our creations will occur more easily too.

Other Allies

As Julia Cameron says in *The Artist's Way*, the one factor with the greatest effect on the success of creative people (and we're all creative people as Brene Brown and many other scientists have proven), is to be part of a supportive group. There are various ways to be part of such a group.

It may grow organically around you, like my 7 year long, weekly lunch group, which began after a Chi Moves class. Over lunches we shared our common interests in wellbeing and cutting edge and traditional wisdoms: what we knew and our newest discoveries. Slowly we became good friends and supported each other. From that many new careers and businesses have emerged, as well as much personal growth.

Or a supporting group may begin more intentionally, such as a work group, class, or social group of mothers etc. Barbara Sher ideas parties are very intentionally focussed, and catalysts for life transformations. And these days they draw in the resources and contacts of many more people because they are run very effectively using Facebook. Barbara's catch-phrase "isolation is the dream-killer" resonates with many people.

Some groups are more effective at supporting your best life than others are. It is said that we become like the five people who we spend the most time with, and current neuroscience and quantum field understandings show there would be much truth in this. In choosing your valuable support groups, go with your intuition.

Find out too about **Nonviolent Communication's** powerful ways of communicating. This is a system or awareness so people can gain better understanding of each other and their needs, and be less counter-productive in relationships. (see next section)

Assertiveness or **gentle assertiveness** replaces fears of being too aggressive. The clearer our boundaries, the more calmly we can express them and maintain them. Less stress. Less upset. Just a clear immovable line (with occasional clear exceptions of specific negotiated "third alternatives" - read on for more on this). Our boundaries are where we have drawn our invisible lines around ourselves of how we wish to be treated, and what we will not tolerate. The clearer we are, the calmer and gentler we can be about ensuring others respect our boundaries. People who are very noisy about boundaries usually are not clear about where theirs are. Assertiveness also helps those who fear confrontation in varying degrees, from a little discomfort to total avoidance at all costs. You can imagine the scenarios that may occur from the second one. To do everything to avoid confrontation means one may be willing to give up rights or may pull right back from relating, or from communicating. Both of these are a loss for the people concerned.

All of these methods can be learned. In fact it would be valuable for some of us to set up circles together to learn about these, and how to include these skills in our lives. Imagine the benefits. Imagine how interesting it could be. Here are some tools to choose from in your group:

- Speakers
- Book clubbing
- YouTube or videos
- Role playing

In "Assume Love", the idea is not to compromise, but instead to find a third alternative that is pleasing to both partners. Perhaps you'll need to broaden your horizons, brainstorm, research, or reach out and ask others for help, but Patty Newbold asserts, with many examples, that there is always another solution to satisfy both people.... a third alternative.

At www.assumelove.com, Patty encourages us to remember that this one person we decided to share our life with, is not planning to do us harm. There are many times that we feel our partner is not responding as we feel they should, or we would prefer, and we can become upset. Instead we honour both of us by assuming that their intention is loving. Just as we'd hope they'd assume that our intentions to them are loving... even when we are very human and perhaps acting childishly. Assume love. They are two powerful words.

Nonviolent communication

"It doesn't really matter what the other person is doing or saying, what their behaviour or words are, everything from them to us is really a version of 'please' or a version of 'thank you'. Because all of us in our own way are seeking to meet needs that we all hold universally."

"The foundation for Nonviolent Communication is empathetic listening: put yourself in the person's place and be curious." I wonder if you're feeling scared?" "I wonder if you're hungry?" "I wonder if you're feeling shame?" And have that empathy to say "I'm not suggesting you're right, I'm not suggesting you're wrong, I'm just wondering what it is that you are needing."

This is a conversation with Fran Westmore[50], trained in NVC, a communication process that helps people to exchange the information necessary to resolve conflicts and differences peacefully.

"NonViolent Communication (NVC) is also known as compassionate communication. It is based on the belief that compassion is our natural state of being, and it's natural for us to want to reach out to other people and want to give and receive from the heart."

"The founder Marshall Rosenberg developed a list of universal needs and universal feelings that all of us share. His approach with nonviolent communication seeks to move out of words and the head, and instead to listen from the heart."

"Marshall made use of some playful things to teach people what he had found. He chose puppets: one of a Jackal and one of a Giraffe. The jackal because it's a creature that is low to the ground, it lives on waste and is very competitive. The jackal represents those times when we are in our head. Where we hear something from

---

[50] Thank you to Fran Westmore, trained in NVC, for this insight. www.franwestmore.com

another person, or from ourselves, and we put our jackal ears on and hear judgment or we analyse or problem solve. We do all of those things that are not actually listening and hearing the other person. We then go into our heads and start to process what it is that is being said."

"The giraffe is the land animal with the largest heart. Also it's very tall, so is able to see beyond the immediate. The giraffe will not analyse or judge, dispute or pre-empt. The giraffe simply looks for the needs and the feelings behind what is being said or done. And consciously seeks those out in an endeavour to make life more wonderful for the other person."

"Marshall told a story about going into a war zone. He managed to gather representatives from both sides, each with generations of deeply entrenched hatred and warring. When he brought them together there was a huge amount of anger. It took several days before they would even speak or be in a room together."

"Once he was able to get them talking to him and then hearing each other, they were able to express the heartache that they had had with their families, and heartache that they'd had from their grief and their loss, and the fear and difficulties of everyday living."

"And the other group discovered that they were exactly the same."

"They were the same. And all of a sudden they weren't enemies, they were the same group."

"There were huge shifts in the ability to care for each other and meet each other and talk together."

"Marshall always said it can take quite a long time to get to the point where you can touch each other's hearts, but when you do the conflict disappears. Because you discover that you are the same person, and you have the same needs."

"Yes" I replied, "that's something that we often don't feel. Often we feel very separate and very different. I can see why this is so powerful"

"Yes and it's immensely powerful in relationship counselling. One of my mentors is a family counsellor, she works together with husbands and wives and families and all the different kinds of partners that we have in the world these days. She uses compassionate conversation with some deeply trenched arguments."

"I can remember one example talking about money. The wife in this relationship had never been allowed to handle the money, and the husband was very dictatorial about maintaining a stranglehold on the budget and the cheque book and bank accounts. This had been the case for the whole married life of over 30 years. They had never been able to resolve this."

"They finally talked about the needs that they had. The wife had always assumed that for her husband it was about control, and she discovered that the need that he actually had was for safety. Because when they were first married she had never managed a budget before and she didn't do it very well because of her inexperience. And they lost some money and it frightened him, he was scared. He was never able to give her back that a quality because he remained too scared."

"He had always assumed that she wanted control of the cheque book so she could be the same inexperienced and irresponsible person with money as she was when they were first married. But what she was actually looking for was respect and trust. Once they realised that, they were able to talk together and they were able to resolve it. Solutions fall out of the conversation once you realise what needs you are actually trying to meet."

"Once you get to the core of it."

"Yes once you get to the heart."

"It sounds so simple, but 30 years without understanding each other and then to have these tools that you can make such a shift in only a day. It is so powerful."

"Yes and I have seen great shifts in people in their own conversations with themselves. Non-violent communication is a very powerful tool in getting past shame if you can have the courage to be vulnerable and tell the gentle truth of yourself, of how you feel and what you need. This takes great courage particularly to begin with. But if you can find that courage and express that truth then the forgiveness of yourself or of other people come."

"So having a NVC practitioner will help to make it safe. So that when one person is trying to be vulnerable, the other person doesn't by mistake come in with big boots and trample on their heart. I guess that's one of the great powers of having someone to walk along this path with you, and help you learn how to do it. You could do it as a couple or as a group."

"Yes you can feel safe yourself once you have practiced enough, and it takes time as it is very alien to the way that most of us have been brought up. The way society looks at the world there is always someone who is wrong and someone who is right. There is always a right action and reaction, always a good decision and a bad decision. A good emotion and a bad emotion. So it takes time to practice and learn how to not fall back into that old pattern."

"But once you do start to do that, you can work with somebody and communicate with somebody, whether they know NVC or not. Because all you ever see are the needs."

"One of the stories that I have heard is of someone who was in London at a NVC conference. Afterwards someone was trying to mug him. He instinctively used NVC with this person who had a knife, and said he was guessing he had to be pretty scared to be doing this. As that encounter unfolded the man said to the mugger "I'm afraid I only have this much money but I'm willing to give it to you". And then "I'm wondering if you'd be willing to hear me tell you that now I'm very scared and anxious because I'm in London and I have no idea how to get back to my hotel because I have no money". And the

mugger gave him the money back. It was because they touched hearts."

"One can only wonder what might have happened next to the mugger. It must have been quite a day for him."

"Yes because encounters where your heart is seen and heard do change you, and who you think you are."

Bargaining

People often will bargain with God or the universe to give up something they love, to help someone they love. Sometimes they will offer to suffer in some way. Again it's a matter of, "If you believe it to be true, then it is".

But the higher truth is that no bargaining is needed. You do not have to give anything up to ask for what you want.

Perhaps sometimes it would be a good thing to give something up that does not benefit you, such as an addiction or a compulsion.

However no bargaining is required. You simply focus on the new reality, and release your counter beliefs.

From stress to creativity

Something has happened to someone you love. They are in an uncomfortable or painful situation. Right now it can't be changed, it just has to be dealt with. We don't want to accept it, but that is the only choice that will give us any relief.

Then we need to work out how we can get back into a creative space rather than sharing their pain and feeling like the victim. This book has a variety of suggestions for this, though one of my favourites is to remember who-you-really-are.... feel the you who is in control.... you are what is often called consciousness or a timeless soul. Suddenly you step out of the pain game, and feel calmer and stronger.

Feeling like a victim removes our power, our peace of mind, and our health. Research has clearly shown stress puts us into fight, flight, freeze or fawn[26]. All of these use all of our energy instead of sending it to our body to maintain, heal and restore itself. These stress responses were intended for short periods of time, to save us from a predator, natural disaster, or perhaps an enemy. If they continue as long term responses, our health and wellbeing will be impaired. Imbalance and sickness will result.

Have I encouraged you to learn more about releasing stress and increasing creativity (which takes you to the very life-enhancing state of flow)? I hope so. There are so many books and YouTube clips where you can find about those two things, and what they teach you will be beneficial. Any changes you make with either of these will have profound effects in your life. You will be increasing the life in your body, mind and soul.

---

[26] Fawn means being nice to others to keep things calm and safe. This fourth one is a relatively new addition to the well-known trio of fight, flight or freeze.

Story sharing and tribe

Sometimes I spend time with virtual strangers, but this can also feel like powerful times of connection, and sometimes the feeling of satisfaction, because I give a gift of myself or my abilities where it's of use.

Recently I spent 3 hours listening to and massaging a woman who was in the middle of months of helping her very small child through a life-threatening illness. She'd stayed with him for months of hospitalisation, invasive tests and treatments. And through it all she and her ex-husband had to deal with each other. Her family weren't emotionally or physically up to dealing with the situation. She was only just beginning to allow herself to ask for help. We sat, she talked. And then, through massage and energy awareness, I helped reconnect her, and helped soothe out the pains and the overwhelming emotions attached to all those physical hurts.

Afterwards we talked a little, as she slowly came out of that delicious between-time of softness and not quite being back in your body.... Both of us had had that sense that we were in each other's tribe - the words had come to both of us. In that state there's a connection. There is unconditional love and understanding, allowing and giving. There are no expectations, as the tribe is a place to be where we are all "better together than apart" as my new friend said.

"The way you see people is the way you treat them, and the way you treat them is what they become"

Johann Wolfgang von Goethe

Checklist: How are your relationships?

How are your relationships with others? Here is a checklist to help you consider. It's worth recording your answers. You may wish to answer separately for each relationship, or each type, such as children, close friends, etc.

1. Are your conversations nonviolent, especially with those close to you? I.e. kind, supportive, positive, enjoyable for you both? Or can they also be manipulative, controlling, cruel, disappointing, worried, fearful or other non-creative approaches?

2. Is your main relationship balanced with independence and interdependence or do one or both of you desperately need the other?

3. Does one of you work to keep the peace? Or is it two whole people coming together to create something even better?

4. Do you laugh or be playful often together? Silly even? Or is the relationship sombre, dark, bored, conflicted or competitive?

5. Do you share the happy times but the downs too when needed?

6. Do you feel an upward movement in the relationship even when dealing with the down side of life? One or both of you feels sad, upset etc., but are you aware of who-you-both-really-are?

7. Do you stand together and look towards some common goals? Or are you isolated?

8. Bonus question: How is your relationship with yourself?

12

Playfulness in your life

Alan Watts once remarked:

> "People suffer only because they take seriously what the gods made for fun"

When you play you don't need someone to describe to you what play is. But for a minute, before we look at play in our lives, let's look at some descriptions of it.

Peter Gray[52] (2009, 2013) distilled play into these 5 points:

1. Play Is Self-Chosen and Self-Directed

2. Play is activity in which means are more valued than ends

3. Play is guided by mental rules which can change, and leave room for creativity

4. Play is imaginative

5. Play involves an alert, active, but relatively non-stressed frame of mind

Children are likely to be absorbed fully in play, more than adults. We adults tend to be playful, or have a playful attitude, while still fulfilling our adult responsibilities. But we can be drawn fully into play, and this will benefit us.

---

[52] Dr. Peter Gray, Curator of Scholarpedia, Boston College, Boston, MA, USA

As we connect more, we naturally play more.

By playing more, we become more connected internally as well as with our natural world and with others.

This is a great upward spiral. Do more of either of them - connecting or playing - and you will find that the other also increases. Greater well-being is the result.

To what degree do we let ourselves be fully absorbed and explore new ways of being, and let go of our current view of how we are? .... Come on, let's explore.

Recipe for a fruitful life: add playfulness

Something changed my life drastically when I was in my 20's. I was attracted by a certain kind of clowning[27], different to what most of us are used to.

No face makeup, just red noses; not clever clowns planning their responses, just clowns who were really in the present like young children.

We were learning to quiet our educated minds, and just to respond from our feelings and senses.

When I first saw clowns playing with people in the street during the Sydney Festival I was captivated, and moved so deeply that I knew my intuition was urging me to learn more.

I was terrified to open myself enough to try to find my own clown, but somehow eventually I took the huge leap into my first weekend workshop. I still feel appreciation for the love and support of two friends who paid for that workshop. With actual twinkles in

---

[27] from the work of Jacques LeCoq and Jan Hamilton

their eyes they presented me with my "clowning scholarship". Julie and Trevor I hope this book reaches you so that you can know how your one act of kindness and faith has opened so many doors for me - I use the awareness from clowning most days of my life! (See Appreciation).... but back to this story.

Play was the main tool we used. We had fun, we tried out new and often silly things in a safe space together. This built trust between us, and encouraged us to further open our hearts, and show more of our wonderful and unique selves.

We could act out our responses to imagined situations, including to make sounds and move our bodies in any way at all. We had the freedom that we allow young children: to play with and explore the edges of who they are and might become.

How long since you gave yourself permission or the opportunity to see what else you might be? Might you be holding yourself within boundaries that you created a long time ago?

There were no rules of what was right, and as long as we felt safe we didn't want to hurt another.

The things I saw in the other clowns! There was so much beauty in every person as they reconnected with their innate brilliance. The hundreds and hundreds of hours that I have watched clowns, or been one, are experiences that are so precious, nurturing, inspiring and illuminating that I can only hint at them with words.

I need to emphasise to you that there is beauty in every person, not just some people. So that definitely includes you.

Clowning really returns you to your openness, to your beauty and to when you were really young, say 2-4 years old.

So many of us adults keep our masks in place to appear strong and capable no matter how we are really feeling.

We don't wish to show the moments that we are nervous, or it hurts deeply, or the moments that are so exciting or joyful. But clowns show all that, and it brings people to laughter or to tears and to so many emotions in between....

It reminds people.

So remember, the more you can show your authentic self, the more it reminds others, and encourages them to show more of their remarkable selves.

Riches occur where play and intuition meet

Gold is available to us, where play and intuition meet. According to Francis Cholle[28] the interplay of these two ingredients is where creativity arises.

Many others have agreed, including Einstein, Leonardo da Vinci and top drama schools. Businesses such as Google now encourage both these aspects in employees, to increase creativity and gain competitive edge through innovative output.

Cholle takes this even further to say that this is also the birth place of our creative genius.

Many have said that each person is a genius in their own way, and that we reach our potential through following our own path, guided by our intuition and by a playful rather than a serious (and usually also stressed) approach.

Linda, an artist, shared two stories from her life with me, where she had been playful and achieved success outside of her usual. I include these because I'm sure some of you will identify with these. Even if you're not a painter, the principle applies to many endeavours.

---

[28] Francis Cholle https://www.youtube.com/watch?v=6pdmfhiV-OU&list=PLaX950GMDgdjREaa6hkgu2Kw3fLWqq9i6 – **TEDx** talk *Intuitive Intelligence - a new approach to solve America's business creativity and sustainability crisis*

The first occasion was an art class where she was painting her version of a Fred Williams scene. She focused on allowing herself to be playful rather than using great effort while trying to replicate the painting. She was surprised and delighted with the result.

Her teacher, a long established artist, was very impressed with her resulting painting, and believed she could sell it for a very good income. The way she had managed to capture the light particularly impressed him, and she had not consciously achieved it.

Can you recall a time yourself when you have been playful in a project and surprised yourself with a great result, perhaps something you have never achieved before?

The second example was with applying for a work position. After spending months applying for literally a thousand positions, and receiving no interest, one day Linda unexpectedly felt a very playful attitude. It was as though she'd passed through it being serious and important. She felt comfortable imagining herself in a position that she'd love and felt she could do well, but had no paper qualifications for.

With a light heartedness and optimism that surprised herself, she applied, and then later attended the interview. You won't be surprised to hear that she was hired, and did the job as well as she'd felt she would.

You can change clouds

You can change clouds. I didn't believe this until my friend Rainbow Butterfly Woman showed me that it was possible. Really she was having conversations with our universe, in a way that she chose.

She would walk on the beach and commune with the goddess spirit which she had felt and occasionally seen there. She always looked so peaceful and happy at this beach, and she was revitalized. She invited us, her friends, there for walks and cloud watching, and began to share the things she saw in the clouds.

Later I began to see these clouds responding to her. I saw their stories change and flow as she told them to us. And when I wasn't there, or when she saw them at home, she would take photos, and show us. She was being validated by the clouds.... a method she chose to focus on as clear exchanges with her universe.

Are we playing with children?

"Do not grow old, no matter how long you live. Never cease to stand like curious children before the Great Mystery into which we were born."   - Albert Einstein

Linda has worked for many years with young children, from playgroups to storytelling, art and movement. I was interested to hear her views on play.

"When I left as teacher of a Steiner kinder I regularly saw the children in passing. They really missed me at the kinder, because I had played with them and given my love and my heart."

Linda's loving face showed sadness and concern as she was telling me about why she let go of another role that she'd loved for most of its 15 years.

Linda: *"The main reason I left 3 year old playtime group was that I couldn't play with them anymore. I had so much paperwork to do that it actually took me away from that wonderful place of just being with them and being able to play. And the children would come to me and say 'Can you come and play with me?' I had to say 'no'. Eventually that hurt too much."*

"After 15 years of running the group, it had changed so that I was required to observe, document and account. I didn't have the time or energy left to do what the children and I both knew to be most precious.... to play together."

R: It's poignant because you feel the power and sadness that they are not getting what they need in a playgroup. But even worse, as a society we have accepted that this is how it is. We are not standing up to correct it."

For 3 year olds you have to do all those assessments. We know those assessments won't help them to develop. It is known that children learn through play. Playing can even heal wounds as deep as PTSD[29] (post-traumatic stress disorder). All of this is validated by various arms of science."

Play and intuition together is the golden area of our lives from which arises our creativity. The children wanted more gold.

Linda: *"We are reducing the quality and quantity of play in children's lives. Even at 3 years old! We want to teach them instead, of us respecting that all humans naturally want to learn about their world, and all have inner impetus to explore and to express who they are."*

"The teachers are stressed and broken, knowing that they are being forced away from helping children to fully develop."

"So what future are we creating? Why are we going backwards with our youngsters from what was working?"

R: Kindergartens have been the educational area which has been the most successful, and they have been based on learning through play, and focussing as well on relationships, belonging, nature and arts.

---

[29] Attachment Play How to solve children's behavior problems with play, laughter, and connection, Aletha J. Solter, Ph.D.,

Linda: "*The only logical reason that I could think that as a culture we would go so backward is that we have become disconnected. So disconnected.*"

"*We have decided that everything must be measured, written, cut up, dissected and seen to be working. And then it might work.*"

And then we hope it will work. We do not trust what we see, feel and know, because we are disconnected from these. We have begun to only trust what we measure. And even when it is obviously not so.

"*In Steiner they say 'children play and they learn through nature. Everything they need to know is in nature. And that's how they learn'.*"

In closing this section, I ask you to look inside yourself to see if there is any truth in this for you. If so don't despair, simply take it as an opportunity for change or growth.

Exploring playfulness

We all have our own unique view of what playfulness is, and we may not be conscious of it.

Some people feel that playfulness is childish rather than childlike. It is however a natural, healthy and essential part of ourselves right through our lives.

I'd like you to explore sometime lately when you've been playful. Again it is a very broad topic and we're going to have a bit of a gentle look at it. I'd like you to think back in the last few weeks when you have been playful.

Let me prompt your memory:

Perhaps you have been having a cuppa with a friend, shooting the breeze in a light hearted way. There's no stress or tension between you, it's just very easy. Perhaps there's some laughter.

Were you outside by yourself, just being with nature? Did you find yourself unexpectedly doing a few dance steps to some music that came on?

Were you playing with a child or an animal?

Or is it some other things that you just naturally do without thought?

That's a few ideas to get you started.

...Now you have your memory, sit with it a little and observe how it feels....Then....

Can you see the list of words entitled P*layful Words* (in the next section)? I'd like you to go through this list and identify emotional states that you identify in yourself during this recent playful experience that you're recalling.

When you're ready we'll continue.

What I want to discuss is: can we be adults who are playful, who don't overthink playfulness, we just relax into it. I am very curious about how comfortable we are with the notion of being playful. Do we allow ourselves to be playful? Should things be serious and have a serious purpose?

Mind you of course, the most powerful way we have ever learnt in our lives has been when we have been playful. When we're playing it is powerful learning and also a powerful connection to our imagination.

I'd love to hear your responses and comments about this exercise. Shae, would you like to start?

At this point there was a mix up with similar sounding names Kay and Shae and light-hearted banter grew into laughter as we played with who might be the greatest over-achiever and how many workshops does one attend, and so on. It grew quite silly, and then we were ready to move on.

Shae: *"The words that jumped out for me, that got my attention, were "silly" and "rude". With the instance I was remembering, we were both being very silly and very rude around the campfire and it was great. It brought lots of feelings."*

"Initially I didn't think I would come up with a negative, because it was a playful experience. And playful experiences are primarily positive. But then I circled the words tired and exhausted because we had a long day's walk. So it's interesting that the two can be juxtaposed and without this exercise I wouldn't have thought of putting those with a playful experience."

R: "Yes thank you very good point. Thank you Shae"

Kay: *"A friend and I were at the movies and we were watching the preview. It was a humorous movie and we were adding comments to it, to make ourselves laugh even more. There weren't a lot of people in the theatre but the people there didn't quite get why we were laughing so hysterically... Like Shae I came up with more than 3 feelings: I felt connected, I felt content, cheerful, happy, relaxed and calm. I didn't feel any of those things that Shae thought of as negative. It was calm and fun and relaxed."*

R: "Yes, you came to it via a different route. Shae had probably been walking all day and was physically exhausted I imagine."

Shae: *"Yes"*

R: "I certainly can relate to both of those situations. I vividly recall being so tired that I was almost delirious with silliness."

Kay: *"And you get slap-happy"* (we all laughed again)

Toolkit for living in an *energetic field* of love

Kelly: *"Me and my man dancing with different styles, different kinds of music. Hyper, joyful, engaged, silly and alert."*

R: *"Aren't these wonderful combinations: hyper, joyful, engaged, silly and alert. You can see why in play there is so much connection to our own power. All these parts are engaged. It's beautiful."*

Kelly: *"For me I had some that weren't even on the list, but I didn't feel capable, graceful or have a sense of rhythm. Those were 3 things that I wish I had have felt."*

Kay: *"Earlier on you asked if adults can be playful. It's funny because some people don't think adults should be. Frequently I've heard from family members that I just look for too much fun. I'll be doing something, it will be a serious something, but I'll make a joke, or take it to an extreme which makes it absurd, and just make it fun so that it's easier to do. Sometimes it's not even something that serious."*

"I know some family members, my brother for example, to get him to joke or to sit there and come up with nonsense is like pulling teeth. And then I have some friends that I just sit and joke with all the time. We can still be very serious and deal with very adult things, but it's not "Oh my gosh I'm an adult I better not smile"."

R : *"Yes the differentiation you are making is very clear. So do you feel that your brother is enjoying his life as much as you are enjoying yours?"*

Kay: *"Oh heck no. That's not even a thought."*

R: *"And enjoyment is such a powerful energy."*

Kay: *"What surprises me is the energy he puts into being angry and annoyed at my having fun. Is he jealous that I'm having fun? What? Because it's not just "You shouldn't be silly like that". It's more of "Why the hell are you laughing? You shouldn't be having fun. This is a serious matter! You shouldn't be joking! We're adults now, you just look like a silly little kid when you say stuff like that. And you're stupid and immature...." He has a lot of energy and anger behind it. And I just don't understand this."

R: "Yes he has a lot of pain there hasn't he?"

Kay: *"I have no idea what's going on with him. I don't know that I want to figure that one out."*

R: "I find mostly when people are angry they have pain beneath it. So for instance they have been made to feel not okay being like that. So number one: they are angry that you're getting away with it when they didn't, and number two: they can also be worried for your survival. If he doesn't shut you down somebody else will trample on you like they did on him. This is mostly all unconscious of course. But anger is usually protecting pain."

Kay: *"That's interesting."*

Words for Playful Exercise

Adamant	Cheerful	Distracted	Flustered
Affectionate	Childish	Divided	Focused
Awed	Child-like	Eager	Foolish
Awkward	Clever	Electrified	Frustrated
Banter	Connected	Energetic	Fearful
Beaming	Conscious	Engaged	Free
Blissful	Content	Enjoying	Full
Bold	Crazy	Envious	Generous
Calm	Delighted	Excited	Glad
Capable	Determined	Exhausted	Good
Captivated	Disconnected	Fascinated	Grateful

Toolkit for living in an *energetic field* of love

Happy	Overwhelmed	Tenacious
Helpful	Panicked	Tense
Helpless	Peaceful	Tentative
High	Pleasant	Tired
Hyper	Pleased	Troubled
Ignored	Proud	Uneasy
Impressed	Raptured	Unsettled
Inspired	Refreshed	Vital
Joyful	Relaxed	Vulnerable
Jumpy	Relieved	Wonderful
Kind	Restless	Weepy
Lazy	Reverent	Wild
Longing	Rewarded	Worried
Loving	Rude	Zany
Low	Satisfied	Zestful
Melancholy	Settled	
Mystical	Silly	
Nervous	Slap-happy	
Nice	Soft	
Nurtured	Solemn	
Nutty	Startled	
Obsessed	Sure	
Odd	Talkative	

13

Creating our New Futures by Feeling and Visioning

"Order is found in things working beneficially together. It is not the forced condition of neatness, tidiness, and straightness all of which are, in design or energy terms, disordered. True order may lie in apparent confusion; it is the acid test of entropic order to test the system for yield. If it consumes energy beyond product, it is in disorder. If it produces energy to or beyond consumption, it is ordered"

- Bill Mollison

This analysis by Bill Mollison, co-founder of Permaculture, turns our society's common view of the value of disorder on its head.

It is time for us to look at things differently, to understand more fully, and design our lives with more powerful tools.

To have a new outcome requires change in ourselves. It may feel different, but it can be a "good different" or a "great different". There are pointers to many kind and satisfying ways to change in this book.

Experience some disorder or even chaos, to achieve a greater yield in your life.

Adding In and Fading Away

In most cases it's so much easier to add something new into your life, and then allow the old habit to fade away from lack of use. This is gentle and kind to yourself. This creates new neural pathways in your brain, and the unused neural pathways disappear. Learning new things, new understandings, new habits, all make connections through your billions of tree-shape neurons.

The messages pass along from one neuron to the next. The more that the occurrence happens, the stronger this pathway of connections becomes. It's like creating a new path through the bush. Often it's hard the first time you create it, but as you use it, and it becomes popular, the more definite and maintained the path becomes.

And on the world level it's the same. Create a new path, don't retaliate. Don't push against a giant system that's not working well for everyone. Begin creating a piece of a system that will work well.... Doesn't that feel better?

There is so much joy in good creation and in great design. There is such diversity of things that appeal to different people. So many positives I could spell out here, but instead I'd like you to create a list. Please join in with Shae, Kelly and Kay as they do this.

EXERCISE: Adding In and Fading Away.

For this exercise, write two things in your life that you have been pushing against. Things in your life that you don't want there anymore, but instead of letting them fade away you have been pushing against them to change.

When I say "pushing against" I mean worrying or thinking about. Perhaps focussing on them with the intention of reducing or stopping them. They might be tiny little things, they might be big things.

You are pushing against or resisting a current situation or reality, or your own or others' behaviours. Behaviours are usually the result of habits, ideas or feelings.

Here's an example: staying up till 2:00 in the morning might be something that I would push against at times.

Shae: *"I will have to write that one down."*

R: "And some things that you rail against out there in the world, you might be very passionate about, such as social justice issues or politics."

So, to repeat the activity: write down two things in your life that you are pushing against. That would be things out in the world, or of yours.... behaviours, beliefs, ideas, habits.

I will wait while you do that. :-)

Hopefully you have found a few and written them down.

Now for the second part.... You now have your two things that you have been pushing against - things that you don't like. Now what are the things in your life that will come in and fill those spaces in your life?

For example, when I don't stay up till 2 a.m. what will my life be like then? What is it that I really want instead of staying up till 2 a.m.? And if you stop pushing against a particular social justice issue, what do you want to do to create something else positive in your world? It might be related to that or it might be different.

So this is a brainstorming session. Just relax and look at your first list, and then see if a new step or a vision pops up - one that you'd prefer. Just write down some keywords which will allow you to remember that whole idea later.

Now you can share as much or as little of your discoveries as you wish to. This is a standard approach of focusing on a positive, but in a short time were you able to come up with something to remind you where you need to focus to create what you want to create? Or did it give you any new Sparks or ideas?

Shae: *"I found it a very helpful question because I think that's the trouble when I'm thinking of the negative, or I'm thinking of what I don't want, it needs a trigger to flick me into what do I want instead. Whilst I already know that intellectually, that exercise was really helpful. So my three triggers are: first: sleep.. Would I be better off sleeping? Because I had exactly that situation: I stayed up and read a whole book until 2 in the morning."* (Recognition laughter from Regina who sometimes does the same.) *"I could easily have spaced it out over the weekend. But no, I had to read it. And that had implications. I had to be up at 6:30 and I was exhausted for my bike riding.*

The second is reflect on what I know versus this continual ongoing search for new stuff on the Internet.

The third one was meaningful work. Continuing to ask is this meaningful work? Or to ask how is this meaningful work? To try and find the connections between what I am currently doing and what I consider meaningful cause-related social and environmental justice areas. A very good activity to get me thinking."

Kelly: *"I got a bit stuck on this actually. I didn't quite get it at first. I was writing things like "as soon as somebody makes the rules I try and figure out how I can break them". It's just the fact that the rules exist. Even though I wouldn't have done the thing even if the rules weren't there.*

And I just can't do anything boring and repetitive, I just can't do it, and don't do it, which I think is why I don't finish sometimes. As soon as I know what I'm doing I am done.

Then a few more simple things.... right now I have a horrible email addiction. I mean it's insane, I check my email way too much. Sometimes I'll be on my computer and I'm on my email on my phone at the same time. Maybe too much iPad and too much TV, those kind of things.

So on the other side when you asked what could I be doing instead, I wrote down pottery, painting, working on books and workshop, so I was writing it's better that I do that. I'm not sure if I understood the exercise."

"At the moment Kelly", I replied, "The things that you do are very real to you. However, it seems to me, that the things you'd like to replace them with don't seem as real. Or they are a big chunk like 'write a book and work on workshops'. Chunks that are so big that you can't see or feel a doorway into them. No starting point. And I'd explore this with you, if that is the case."

Kelly made it clear that she was interested to know more, so I continued.

"With 'write a book' you might need to break that down into small steps, so that when you need something to do you can go towards something specific that you have already envisioned. These might be a few of those steps: 'Think of a great name for my book', 'Test my book name on others', 'Trawl the internet for a book cover designer'. We'd be meshing them into what you are doing already, and edging you across to the new world where you want to go. "

"At the moment I can't see why you would go from where you are to what you want, because it doesn't sound real or appealing at this point. But with a bit more brainstorming and a bit more playing I think you could find things that you would want to add in."

Later on, after a playfulness exercise and further light-hearted conversations, Kelly had a buoyant AH-HA moment. It was such a delight to watch the transformation. Suddenly she could see herself using her talents in graphic design, to create a more fun, beautifully laid out book than her earlier design idea. It would have much less content, and be a better fit with her 20's and 30's target market. In this case less is not only more, but she was suddenly alive again, with eagerness to start to lay out the book. It just took some time to check in and find a way to direct her project so the joy and creative energy would flow.

Kay: *"Some of the things that I do, or that are going on now that I don't want: I see myself 'giving in' too easily to demands from others; I also stay up way too late. I'm a night owl so it's a natural thing for me to do, but it's getting out of hand. I spend too much time surfing the internet, including late at night. Another one is not taking enough action.*

So then I was looking at what I replace or what I get by not doing those things. Some of them are just a little too broad, like giving in too easily. I shouldn't give in so easily because then I don't get the things I want as much. But that's kind of nebulous so I need to see how to chunk that down.

Staying up way too late surfing the internet: I love to read and I have bunches of books that I want to read so why don't I just grab a book?

Not taking action: Frequently I will be sitting here for 8 or 9 hours at the computer, I will get up to go to the restroom or get lunch and the rest of the time I will sit. That's more time at the computer than when I had a full time job. It's ridiculous! And at least when I had my full time job I was making time in the evening to go putter in the yard or go to a park or something like that. But I need to find something that will make that easier, because I regret it later when it's too late to go.

This morning was different, I thought "I need to tackle that pile of wood in the yard," and I don't know what the trigger was, but suddenly I said to myself "just put your clothes on and go and do it already!" and I did! I don't know what made that trigger actually work."

R: "So you're saying that somehow you stopped and noticed?"

Kay: "*Yes*"

R: "You often continue on without stopping to reflect, but in this circumstance somehow you did stop and have a conversation with yourself. Is that what you're saying?"

Kay: "*Actually I always know what's on my calendar, and there was nothing until that afternoon. So when I woke up I spoke to myself and said "instead of going to the email to see if there's something I need to answer, I can go out in the yard for several hours and that email isn't going to rush away." I had breakfast and I didn't even allow myself to turn on the computer. It was like: get up... eat... clothes on right away... and get out.*"

R: "Ok, so this is a very good example. Really you created something quite clear and different to move towards. You talked yourself in a new direction. You consciously focused and chose your intention."

Kay: "*I need to do it more often.*"

R: "Good, so after this call, will you just focus again on how that happened? Relive it as though you are there. And then relive it as happening in your future, and still as real."

Think about how good this choice feels. Allow yourself to feel the satisfaction and the joy of this choice, and get yourself excited about the next chance to do it again. Start to make 'adding in' more real and more enjoyable.

For example, some questions to ask: What other techniques could I use? I could meet a friend next week, I could put that in my calendar. Wherever you find that you are doing things that you don't want for long periods, how can you inject a change point? As you've said, things that are on your calendar will move you.

How about you stop now, plan something and take a step to make it happen? Ask yourself what you would love on your calendar. Then take that small step: a phone call or message; or put it on your calendar. Now is always the best time.

Do you notice any small change in how you feel now there is another step organised in your new direction? Your neurons are building new pathways that are creating the life you want. Keep on with this and soon you will have your own new forest of neurons, and your life will have more sparkle.... Remember.... Kindness Is Small Steps <3

It's very interesting to be doing this workshop with people who have similar habits.... For me, if other people aren't involved I might just rebel and stay rooted to the computer, or not even think about moving for a whole day. To ensure I move my sometimes neglected body, my best tactic is to organise meeting a friend for a walk, or helping each other in our gardens or to meet up at a class.

Kay: *"Now this is kind of bizarre, but I'm finding I'm having to tap my anger, or get myself angry, so that I'll protect my boundaries and not give in to suggestions that people make, or not giving in to whining by my brother. I really have to have a self-talk.... such as "We are going to get together, and somebody will make a suggestion about the restaurant and if there's even a whiff of you not liking it, then open your damn mouth!" Or else I'll go along with things, and I find myself at another restaurant that I don't like and I'll be wondering why I agreed to go there."*

R: "When you are talking about talking to yourself, you are describing it as being angry and I am hearing it as just having a clear boundary. You're having a strong chat to yourself to remind yourself to keep your boundaries strong. That's what I hear."

Kay: "*Oh!*"

R: "And it sounded like you needed a pep talk, like before a game, to remember to support yourself. Because there will be people there who are pushy, so you can't be your normal gentle self with them, you need to be tougher."

Kay: "*I have to put my foot down.*"

R: "Yes, you have to make sure you say what you need. But I didn't hear any angry talk. I heard "I have my boundaries up and I can do that. I do have to be vigilant. I definitely need strong boundaries around these people. Otherwise I'm going to get clobbered again."

Kay: *"Thank you for showing that to me in a different light. That's good."*

A short exercise on NEWNESS

First: Stop. Observe your body. Then gently take 3 slow, deep breaths.

Now before you read each question, stay as relaxed as you can. Be as open as you can to an answer occurring to you. Accept whatever it is. Write it down in your notebook. And keep breathing.

Here are your questions, for finding the speed at which you can comfortably add new things into your life:

1. How often is it comfortable for you to make a change, in particular to add something new into your life?

2. Would once per season, which is 4 new additions per year, suit you?

3. Could you add something every 2 weeks?

4. Would you be comfortable sometimes focusing on perhaps 3 new related projects? Or is one at a time enough?

Your answers are written somewhere? Ok, let's continue.

If we add things too fast it becomes overwhelming, tiring and they don't stay. In permaculture "long and protracted observation and thought" are keys to good design. Equally, regarding newness, it is very important to observe ourselves and our environments and understand what works for us. (See Feeling yucky is just a habit)

Visioning to make it so

"Divine Intelligence loves to work for you. It is your obedient servant that awaits your recognition and command. Scientists state that you live in a sea of intelligence which is moved upon by your thoughts[30]"

Catherine Ponder

As we are complex, there are many reasons why something may occur in your life, and it is likely to be a combination of things. However action in any of the areas will move you forward, and may well take you to the next action to achieve your goal.

So remember, for all dis-ease, for anything unwanted, whatever it is, create a vision of how you choose to be, and start to feel and act as though you are already there. For in truth you are a piece of the divine - a creator in a participatory universe, connected to all. It's just that we are programmed to keep forgetting this. So kindly remind yourself AGAIN as you feel your powers. No wonder we all love Super-Heroes, because that's who we all are! Really we are like Clark Kent, forgetting he's Superman, or like Diana Prince, forgetting she's Wonder Woman.

Release any doubts that appear. You are setting a clear intention, and creating it. From here help, guidance and actions will flow.

Be kind to yourself as you learn to use your powers and make inner changes.

When your doubts keep showing up, be kind to yourself but firm and persistent in using new techniques such as affirmative self-talk. (See Brain washing)

---

[30] The Prosperity Secrets of the Ages, Catherine Ponder, DeVorss & Company, 1964, Revised 1986

Living in your desired life

It's relaxation and creation time again:

1. Sit or lie down comfortably.
2. Focus on your breath as you relax.
3. For any worries you have, personal or about others - choose one.
4. Create a great outcome for your worry.
5. See/feel this play out in your imagination, either static or dynamic, (still or moving).
6. Make it feel real. Involve your senses.... what you can see, smell, hear, taste, and touch, including sensations in your body, and emotions.
7. Especially have your own higher self present, and that of the others. It's that part of us all that is strong, loving, secure. It's the pilot that is in control.
8. You are creating your future.
9. The more that you can re-create and feel this oasis of calm, the better you'll go. (and like anything else, it will become easier).

Temporarily Lost on the Journey

Lost in a dark, cold, damp forest

Can't hear or be heard. So incredibly leaden and lonely

I can't see. There's no point. No vision. Nothing makes sense.

It's senseless! I'm senseless too.

How am I blindfolded?

Earplugs in, or tuned to someone else's output?
Wet heavy blankets on me, blocking touch.
Denying myself physical experiences, not living my potential.
Chastity belt, or other kinds of locks? Where are the keys?

Censoring my wisdom, truth, or my singing voice?
Surely I have some value?
Then why not inhale and taste the joys of life?
If I'm not making choices that support and nurture my senses
I haven't fully "come to my senses" yet. I've backed away.
(and probably for basic survival or protection)

The journey goes on and fire clears the debris of a life part-lived.
Burn-out. Nothing left - no fire, passion, or drive.
Black grief. All of that for nothing! Stuck in nothing!
Only ashes of self left - that's the belief - it's not TRUTH!
But am so disconnected from senses and world that it seems true.

Eventually I found proof -in a million scientific papers -
The World is Truly Conscious and Responsive.
All of it!!
Just imagine....

This is so hard to let sink in...
I am a co-creator.
What I do affects everything,
Everything else affects me.

So have I helped?

Or have I mucked up the world?

Hang on.... would I blame a baby for falling over while trying to walk?

Or would I cheer, and share the joy of each step?

We are the babies, small in such an immense

and fascinating world of possibilities.

We create.

....So surely it's our fault? (My inner voice will not give up)

NO!! ...We've done our best. All of us. Believe it or not.

It's about growth and the journey of coming to our best.

It's a Journey. To. Love.

Coming to love, and loving the journey.

There's no blame, just signs to read.

We truly don't need wars, cruelty or shame.

Expanding,

Light-hearted,

Curious.

Joyful visions of a Great World.

Helping to birth new ways.

Creatively using our connecting web,

With friends, allies and infinite choice....

When you're not the meditating kind

Is meditation a state to achieve? Or is it our natural state when we stop? Let's look into this.

One can be in a meditation or alpha state in many situations that you may not have considered.

1. Being massaged.
2. Eating ice-cream or a good meal.
3. Involved in any of your passions including arts and creation.
4. Making love.
5. Being with family and friends - again the love factor.
6. Being in the flow....
7. Daydreaming.
8. Being in nature.

Formal meditation and I have had a tentative relationship. It's not too terrible really, but I've found that it isn't always my best method - sometimes it works and sometimes I feel like I drift off and before I know it I've organised my shopping list and errands, and been thinking about other people and various challenges. Not necessarily a respite. I tend to forget it or just do it for a short time.

On the other hand there are many other things that I have done that put me in a meditative mood.... and I have been assured of the validity of this by more than one meditation teacher. See above for a list of some of them.

I'd like you to consider: is a meditative state something that we are "trying to achieve"? Or is it our native state which we return to naturally when we leave a bit of space in our action-paced lives?

Is it our soul or spirit that is in connection with all, when we allow ourselves the chance to sense and feel it?

I guess you can tell I believe it's the second one.

Small Steps ARE Powerful with Sue

"So let's see.... I can apply many of them", says Sue, considering the list of ideas I'd given her.

Sue is stuck in the midst of her new project. Something that she wants to do, but each time she intends to work on it she finds herself "bouncing off" and finding many ways to avoid her desk. Sue is very keen for this to be the basis of a new business, for which she has the right qualifications and experience. But she often feels so uncomfortable, and sometimes she catches herself worrying and actually denying her abilities. It's "What's the point?" and "This work won't help people", and I imagine she hits the basic one of "I'm not good enough". She was so frustrated with herself!

Here is the introduction to the list I gave her:

Checklist: When Resistance is slowing you or stopping your action.

....Warning, all of these will have side effects which will include: less stress, more joy, greater peace of mind, and no drugs required.

For all of these exercises that follow, just do one of them for 1-5 minutes immediately before you start work. No further analysis. Do not stop to argue.

Each time you feel any negativity arise to block the flow of your work, either write the words down (blurts) to be dealt with later, or do another of these brief exercises.

Here are Sue's comments about how useful she expects them to be for her:

Jump for Joy (*yes I can do the jumping - oh what a feeling aka Toyota*)

Laughter, including Laughter Yoga (*I hate fake laughter so I will probably avoid this but I could find some of my favourite cartoons e.g. Leunig which often give me a laugh*)

One minute of peace (*potentially yes but it's a concept - probably easier to do the breathing*)

Always focus on the result as already being real (*potentially yes but not sure exactly how to do this*)

Could listen to your voice recording you've made of this, for a few minutes, or look at your vision board. Don't allow space for the negative *(OK yes)*

Exchange sincere compliments with a friend before you write (*yes I like this and I like that you have taken the time to write this out for me - thanks!*)

Read notes of thanks from happy clients(*I have a brag file so I can go look there, but when I'm feeling ugggh nothing comes easily to mind*) - ok, put brag file on desk.

Recall and relive a moment of great joy in your life *(as per above)*

For one minute write, nonstop, what you are grateful for (*as long as I use the word appreciation this one is the one I like the most*)

Spend two minutes planning or feeling your next holiday, or a major one that's not the next *(not relevant)*

Write yourself a love note *(hmmm possible)*

Be your own mentor and have a loving appreciation session of YOU (*nice - a different take on what I usually do - appreciating the world*)

Turn on some dancing music that you love and MOVE it! (*Yes - I need a few fave songs on my laptop not on the iPod in the lounge room*)

Stand in the Wonder Woman power stance *(yes)*

Stand in a grounding yoga pose *(possible)*

Shake every part of you, perhaps like a dingly dangly scarecrow *(possible)*

Do one or two moves from tai chi, shibachi, qi gung, or other *(possible)*

Take three long, slow breaths, allowing tension to release, and focus*(yes)*

After some discussion Sue shares how she'll begin to implement these. "*When I wanted to swim later in the season and the water was getting colder I gave myself permission to have an "easing into the water routine" where I would water walk for a minute or so, then breaststroke for a minute or so then start putting my head under. I think I will apply the idea of a routine - breathing, smiling, wonder woman pose and writing about what I appreciate - for a few minutes or so before starting my work session. That feels good. Gets three of the senses involved.*"

I acknowledge Sue's plan. A plan is half the battle, and remembering and being willing to carry it out is the next half. The better you feel about it the more likely you are to enact it IF YOU REMEMBER. So how to remember? Here are a few ideas to prime your own (with spaces for you to fill in):

1. Put on phone or desk scheduler or alarm

2. Write or draw note and leave on top of your work

3. Put note on kettle

4.

5.

This is all part of lovingly reprogramming yourself. So here are Sue's exercises, conveniently listed for you to print out.

For all of these exercises that follow, just do one of them for 1-5 minutes immediately before you start work.

No further analysis.

Do not stop to argue.

Small steps ARE powerful Checklist

1. Jump for Joy

2. Laughter and Laughter Yoga

3. Take three long, slow breaths, allowing tension to release, and focus only on your breaths and your body

4. One minute of peace

5. Always focus on the result as already being real. Could listen to your voice recording you have made of this, for a few minutes, or look at your vision board. Don't allow space for the negative.

6. Exchange sincere compliments with a friend before you write

7. Read notes of thanks from happy clients

8. Recall and relive a moment of great joy in your life

9. For one minute write, non-stop, of what you are grateful for

10. Spend two minutes planning or feeling your next holiday, or a major one that's not the next.

11. Write yourself a love note

12. Be your own mentor and have a loving appreciation session of YOU

13. Turn on some dancing music that you love and MOVE it!

14. Stand in the Wonder Woman power stance.. as Sharon said

15. Shake every part of you, perhaps like a dingly dangly scarecrow

16. Stand in a grounding yoga pose

17. Do one or two moves from tai chi, shibachi, qi gung, or other.

Living in your desired life

"Imagine what seven billion humans could accomplish if we all loved and respected each other. Imagine."

Anthony Douglas Williams

Thank you so much for coming on this journey with me. My hope is that you have found one or more tools of value to you. If so, please put them in your tool belt as you move on, and spread the word.

Remember we are all connected in this world, and even a small number of people working together can make great changes.

I encourage you to enjoy the adventure of exploring new paths in the wonders of our infinite universe.

It is not surprising to me that Rachel Carson and Bill Mollison, both with a strong environmental consciousness, offer what I consider to be very sane solutions.

I will let their words be the last in this final chapter.

"A people without an agreed-upon common basis to their actions is neither a community nor a nation. A people with a common ethic is a nation wherever they live. Thus, the place of habitation is secondary to a shared belief in the establishment of an harmonious world community.

Just as we can select a global range of plants for a garden, we can select from all extant ethics and beliefs those elements that we see to be sustainable, useful, and beneficial to life and to our community."

- Bill Mollison

"The human race is challenged more than ever before to demonstrate our mastery, not over nature but of ourselves."

- Rachel Carson

Bibliography

This is a very brief bibliography, though with the footnotes throughout the book, it is enough links to keep you engaged for a year or three. These are some of my favourites. In many cases I list just the author or organisation as they are prolific, including YouTube clips.

HeartMath Institute: many tools, including digital and feedback tools, for learning these "new" skills, for children, teenagers, families and adults.

IONS Institute of Noetic Sciences http://noetic.org

The Global Consciousness Project, Meaningful Correlations in Random Data

http://noosphere.princeton.edu/

The Intention Experiment: Using Your Thoughts to Change Your Life and the World, Lynne McTaggart

Alan Watts

Bruce Lipton

Candace Pert, Ph.D on Miraculous Healings

Donna Eden

Doreen Virtue

Dr Joe Dispenza

Gregg Braden

Hay House Radio

Joseph Campbell

Laura Koniver, MD combines intuition with conventional medicine for deeper healing of the underlying energy dynamic.

Dr Marilyn Shlitz

Morty Lefkoe & Matt Riemann The Power of Belief

Dan Millman, "The Way of the Peaceful Warrior"

Pamela Harrison http://www.thenextstepcoaching.com.au/

Chart of Nutritional deficiencies that are indicated by your body, by PositiveMed http://positivemed.com/2013/05/20/detecting-nutritional-deficiencies

Article: IONS Scientist Publishes Cutting-Edge Textbook on the Science of Consciousness: Transcendent Mind: Rethinking the Science of Consciousness http://www.noetic.org/blog/communications-team/ions-scientist-publishes

Local to Melbourne's Mornington Peninsula area

Sacred Soul Healing
https://www.facebook.com/sacredsoulhealing/?fref=ts

Chi Moves www.chimoves.com.au

Healing Station, for healing and spirit art
https://www.facebook.com/healingstationSeaford/?fref=ts

Body Harmony https://www.facebook.com/Body-Harmony-Moves-by-Netta-352175044977526/?fref=ts

Clip
What causes depression John Gray
https://www.youtube.com/watch?v=JtCiA95dzHI

Movie
"What the Bleep Do We Know?" writer, producer, & director Will Arntz

Acknowledgements

My women friends have been a special gift in my life and on my journey. Thank you all for sharing your depths and discoveries with me: laughing together and at ourselves; being playful; vulnerable; lost and wise. Your insatiable appetites for life, and for so many aspects of how our universe works - these have helped inspire, clarify and shape this book. Michelle Tolley, Karen Ann, Fran Westmore, Josie B, Netta Ditchburn, Linda D'Silva, Michelle Endersby, Megan Sinclair, Julie, Maureen Griffin, Chris Highgate, Carolyn Ketels, Kim Isaacs, Jane A. Also to my Artist's Way group where the intention for this book was set: Amy, Leeann and Sue.

All my family and my "children" for all your support and love, and all the lessons we help each other to learn.

There are so many more men, women and children than I can include here, but my appreciation of your presence in my life is timeless.

I have a new group of women to appreciate.. existing and emerging authors... led by the amazingly creative Barbara Sher who not only worked out how to write a New York Times Best Seller, but how to work with others to improve their lives over the last 40 years. Thank you Barbara, and Patty Newbold for sharing so many of your insights and methods with us. This has been an extraordinary year of learning, creating and collaborating with a team - three of my very favourite things. My deep appreciation to all who have helped me to birth this book.

In the final intense weeks, helping me to quickly jump out of pot-holes, real and virtual, I wish to mention the outstanding creativity of Fran Westmore, Linda D'Sylva, Karen Ann and fellow writer Sharon McGann.

The Author

Regina Orchard brings what she has learned from her lifetime of exploration. Clear and from the heart, her headline is...

Science now proves the truth of who-we-really-are, which allows **us to perform what most of us would consider miracles.**

In Regina's words: *Life is an adventure, and it's much more fun when you know how it **really** works.*

Her early experience of academic science felt an insufficient explanation of the fullness of life. This led her to years spent exploring widely, the crafts of clowning, wellness and natural healing with people as well as our land, and a discovery of self-growth as a universal need and purpose for all beings.

Regina, a proud mother and grandmother, and as a multi-potential has combined a breadth of training and work in Computer programming, Sustainability, Online training, Wellness, Massage , Performance Arts Therapy and Community building, with a depth of life experience and deep reflection.

With a message that joins old knowledge and new, her greatest desire is to assist in the evolution of empowered people creating a loving, sharing and sustainable world.

Where to from here?

You've read the book, or parts of it. I hope it has brought you what you need for your journey.... Or perhaps has set you on new paths.

It would be wonderful if you choose to send me a sentence or two about you and this book (for publication) you can leave any messages on my website.

If you would like to continue your journey with me, over time, on my website **www.ReginaOrchard.com** you will find:

- Shop for this book and e-books and audio books
- Workshops
- Personal sessions
- Online gatherings
- Many more links
- Gifts
- Mailing list to hear the latest good things

Or talk to me about facilitating a workshop or talk in your town, or any other ideas. (Connections, a great future, playfulness and expressive exercises are my special loves.)

I am grateful for your support, and for the connection we have made together.... I look forward to us helping each other and helping the world, with as much ease and joy as possible.

Namaste (the spirit in me honours the spirit in you)

www.ingramcontent.com/pod-product-compliance
Lightning Source LLC
Chambersburg PA
CBHW070607300426
44113CB00010B/1444